Biomedical Natural Language Processing

Natural Language Processing (NLP)

The scope of NLP ranges from theoretical Computational Linguistics topics to highly practical Language Technology topics. The focus of the series is on new results in NLP and modern alternative theories and methodologies.

For an overview of all books published in this series, please see *http://benjamins.com/catalog/nlp*

Volume 11

Biomedical Natural Language Processing
by Kevin Bretonnel Cohen and Dina Demner-Fushman

Biomedical Natural Language Processing

Kevin Bretonnel Cohen
University of Colorado, School of Medicine

Dina Demner-Fushman
National Library of Medicine

John Benjamins Publishing Company
Amsterdam / Philadelphia

Library of Congress Cataloging-in-Publication Data

Biomedical Natural Language Processing / Kevin Bretonnel Cohen and Dina Demner-Fushman.

p. cm. (Natural Language Processing, ISSN 1567-8202 ; v. 11)

Includes bibliographical references and index.

1. Computational linguistics--Statistical methods. 2. Natural language processing (Computer science) 3. Biometry. 4. Medical statistics. I. Cohen, Kevin Bretonnel, editor of compilation. II. Demner-Fushman, Dina, editor of compilation.

P98.5.S83B56 2013

006.3'5--dc23 2013029704

ISBN 978 90 272 4997 5 (Hb ; alk. paper)

ISBN 978 90 272 4998 2 (Pb ; alk. paper)

ISBN 978 90 272 7106 8 (Eb)

John Benjamins Publishing Co. · P.O. Box 36224 · 1020 ME Amsterdam · The Netherlands

John Benjamins North America · P.O. Box 27519 · Philadelphia PA 19118-0519 · USA

Acknowledgments

Our thanks go to our colleagues whose innovative biomedical NLP research provided the foundations for this book and who reviewed some or all of it. Here is an incomplete list of those whose work made this book possible: Sophia Ananiadou, Alan Aronson, Christian Blaschke, Wendy Chapman, Mark Craven, Laurie Damianos, Marcelo Fiszman, Noemie Elhadad, John Ely, Carol Friedman, Ken Fukuda, Bob Futrelle, Aaron Gabow, Rob Gaizauskas, Graciela Gonzalez, Jorg Hakenberg, Marti Hearst, Larry Hunter, Helen Johnson, Cem Kaner, Judith Klavans, Zhiyong Lu, Kathy McKeown, Preslav Nakov, John Pestian, Thomas Rindflesch, Larry Smith, Padmini Srinivasan, Junichi Tsujii, Alfonso Valencia, Karin Verspoor, Ellen Voorhees, John Wilbur, Pete White, Hong Yu, and Pierre Zweigenbaum.

We are very grateful to Matthew Simpson and Ivo Georgiev for their thorough and thoughtful reviews of the book.

Table of contents

List of figures

CHAPTER 1

Introduction to natural language processing

As the intended audience for this book is natural language processing specialists who want to move into the biomedical domain, this chapter provides only an overview of the field. Computational biologists who would like a more solid background in the fundamentals of natural language processing and computational linguistics than is provided in this chapter should consult Jurafsky & Martin (2008). However, the chapter provides background on some of the basic issues, and contains information on natural language processing fundamentals that are specific to the biomedical domain that should prove helpful to experienced natural language processing specialists, as well.

1.1 Some definitions

Natural language processing, *text mining*, and *computational linguistics* are frequently used as synonyms. However, from a research perspective, it is useful to be aware of differences in the kinds of questions that are asked and the kinds of answers that are produced by each field.

1.1.1 Computational linguistics

Computational linguistics per se deals with developing computationally testable models of the human language faculty. Questions in computational linguistics research typically deal with some aspect of that faculty, such as the minimum necessary grammatical power for describing natural language syntax. Their answer is typically formal.

1.1.2 Natural language processing

Natural language processing is a subfield of linguistics and computer science that deals with computer applications whose input is natural language. Natural language processing research typically deals with building applications that process language at some given level of linguistic structure. For example, a natural language processing application might be built to tag the parts of speech of words. The answer to a "question" in natural language processing research is typically a value for some performance metric.

1.1.3 Text mining

Text mining is the production of applications that perform specific tasks to meet information needs or provide services. A text mining application might use a natural language processing application for labelling parts of speech to extract statements about protein–protein interactions from scientific journal articles.

1.1.4 Usage of these definitions in practice

In practice, these terms are often used interchangeably, and there is significant cross-talk between the three research communities. For example, the annual meeting of the Association for Computational Linguistics is full of papers on natural language processing, and in the biomedical field, the terms *natural language processing* and *text mining* are typically used synonymously.

1.2 Levels of document and linguistic structure and their relationship to natural language processing

1.2.1 Document structure

Most research in natural language processing deals with newswire text, which has a relatively simple document structure. Document structure is a more complicated issue when dealing with biomedical text. The two main genres of interest in biomedical natural language processing are scientific journal articles and clinical documents. Each presents its own set of challenges.

Scientific journal articles are typically separated into an abstract and an article body. To date, the vast majority of research on journal articles has dealt with abstracts. However, in recent years, there has been a move towards more work on full-text articles, i.e. including the article bodies. This has introduced a new set of challenges. The structure and content of abstracts and journal articles are demonstrably different (Cohen *et al.* 2010a). Structurally, article bodies contain longer sentences than abstracts, and make much heavier use of parenthesized material. This parenthesized material presents a number of problems for natural language processing applications (Cohen, Christiansen, & Hunter 2011). On a linguistic level, there are statistically significant differences in the incidence of passivization, negation, and pronominal anaphora (words like *it*) between abstracts and article bodies, with article bodies showing greater use of passivization and negation and abstracts showing higher use of pronominal anaphora. The density of semantic classes varies between the two, as well; for example, mutations are mentioned far more frequently in article bodies than in abstracts, while drugs and diseases are mentioned somewhat more frequently in abstracts than in article bodies. These differences between abstracts and article bodies correspond to differences in the performance of extant natural language processing tools. For example, part of speech taggers perform statistically significantly better on abstracts than on article bodies, and gene mention recognizers perform worse on article bodies – sometimes drastically worse.

Article bodies are typically separated into sections, with a common set of sections being an introduction, a materials and methods section, a results section, and a discussion section. Materials and methods sections are notorious sources of false positives for many types of applications, and they are often omitted from processing in system evaluations, but the ability to handle them is crucial for extracting the information on methods and on biological context that biologists find crucial in interpreting experiments and the output of text mining applications.

Although article bodies present a number of additional challenges as compared to abstracts, there is abundant evidence that the ability to process them will be crucial to reaching the full potential of biomedical text mining, and they are increasingly the subject of biomedical natural language processing research (Garten & Altman 2009; Lin 2009; Agarwal & Yu 2009; Czarnecki *et al.* 2012).

Clinical documents are typically structured, but there is an enormous amount of variability in this structure between different types of documents (e.g. discharge summaries versus nursing notes), of which there is a myriad. Document structures for the same type of document, e.g. discharge summaries, may vary from hospital to hospital, from department to department within the same hospital, and even from physician to physician within the same department. The ability to segment clinical documents into sections and to label the sections appropriately is crucial. For example, in determining a patient's current problems, it is crucial to differentiate between the history section and the diagnosis section. A small number of tools for segmenting clinical documents exists, but much more research is required in this area, and clinical natural language processing practitioners will often find the construction of an ad hoc document segmenter to be an unavoidable first step in building an application. Demner-Fushman *et al.* (2011) describes an iterative procedure for building such a segmenter for a range of clinical document types.

1.2.2 Sentences

After segmenting the document, sentence segmentation, i.e. finding the boundaries between sentences, is frequently the next step. Doing this for newswire text is approximately as difficult as a homework problem. However, biomedical text presents additional difficulties. For example, sentences in biomedical journal articles can begin with a lower-case letter if they begin with the name of a gene and a mutant form of the gene is being discussed, for example

> eya-clones de-repress hth.

In clinical documents, the situation is even more difficult – in clinical text, it may not be apparent what constitutes a sentence at all. For example, x-ray reports, one of the earliest subjects of medical natural language processing research, are replete with text like

> Clear lungs. No evidence of focal pneumonia.

This text clearly contains two separate textual segments. Each makes a separate assertion. However, neither is typical of an English sentence, both lacking verbs, for instance. Clinical text often also features lists consisting of items that are semantically and syntactically discrete but lack any punctuation, e.g. "problem lists" and lists of current medications:

> Active Problem List
> Erythema multiforma SJ syndrome
> Acute renal failure
> Transaminitis
> Chronic pain secondary to oral lesions
> Right arm DVT
> Chronic Problem List
> hx strep infection

1.2.3 Tokens

Tokens are individual elements in the text. These include discrete words, but also punctuation marks, which in most cases must be separated from the words to which they are attached. Tokenization may also include removal of affixes like *n't*, genitive markers, and the like, depending on the design of the application. In biomedical text, as in other textual genres, tokenization can be problematic. For example, in journal articles, it is generally the case that hyphenated words are kept as a single unit. However, there are cases where the correct semantic interpretation requires splitting them. For example, in a journal article, *freac1-freac7* has the interpretation *freac1, freac2, freac3*, etc., through *freac7*, and the sequence *freac1-freac7* should probably be split into the three tokens *freac1 – freac7*. The sequence *IL3-5* has the interpretation *IL3* and *IL5*, and again, should probably be split into three tokens. Clinical text presents its own set of tokenization challenges and associated difficulties of semantic interpretation. For example, *–fever* has the interpretation *no fever present*, and the hyphen and word should probably be separated into two tokens.

1.2.4 Stems and lemmata

For some applications it is useful to reduce words to their "stems." *Stem* is a term with no linguistic definition; it is purely an artifact of natural language processing. It is a normalized form of a word without inflectional and sometimes without derivational morphemes that is produced by some natural language processing application, which might not necessarily correspond to any form in the actual language. For example, for the inputs *phosphorylate, phosphorylates, phosphorylated, phosphorylating*, and *phosphorylation*, a commonly used stemmer produces the form *phosphoryl*. Note that this is not an actual English word and does not even accord with our intuitions about what the base of the words might be, i.e. *phosphorylat*. However, stemmers can be useful as features for machine learning applications and as components of rules for rule-based systems.

In contrast to a stem, a lemma has a clear linguistic definition – actually, two. According to one definition, the lemma of a word is the set of inflectional forms of that word. For example, the lemma of *translocate* would be *translocate, translocates, translocated, translocating*. On the other definition, a lemma is the base form of a set of inflected words, such that the lemma of *translocate, translocates, translocated, translocating* would be *translocate*. Lemmatization is a more difficult task than stemming, and is much less often attempted. A tool specialized for biomedical journal articles is the BioLemmatizer (Liu *et al.* 2013).

1.2.5 Part of speech

Lexical categories, more commonly known as *parts of speech*, are categories that determine the inflectional morphemes (e.g. plurals for nouns, past tense for verbs, etc.) that a word can take and the slots that it can fill in grammatical structures (e.g., an adjective can appear between *the* and a noun). Natural language processing systems typically assume about eighty parts of speech. We get from the eight parts of speech of traditional English grammar to the eighty parts of speech of natural language processing by subdividing the traditional ones. For example, singular nouns and plural nouns are considered to be two separate parts of speech. Part of speech taggers require retraining to work well on biomedical text, whether it is journal articles or clinical text.

1.2.6 Syntactic structure

Syntactic parsers generally require retraining to work well on biomedical journal articles. Nonetheless, the syntactic structures of biomedical journal articles are probably similar to the structures of scientific publications in general, which have been studied by Biber *et al.* (1999). A number of approaches to simplifying the parsing problem in biomedical text have been taken. A common one is to "detokenize" some semantic or syntactic class of entity, e.g. by finding gene names like *breast cancer associated 1* and turning them into a single token *breast_cancer_associated_1*. Another common approach is to delete parenthesized material, although Cohen, Christiansen, & Hunter (2011) demonstrated that this is sometimes informationally lossy.

1.2.7 Semantics

Whereas lexical semantics is still understudied in biomedical text, semantics is widely used as synonym of domain knowledge. The biomedical domain is one of the few domains with very rich machine readable sources of domain knowledge captured in databases and ontologies, discussed in detail in Chapter 7.

CHAPTER 2

Historical background

This chapter will provide a brief history of the field, from early medical work in the 1960s through genomic work in the late 2000s. A brief review of available resources (lexical resources, tools, etc.) will cover those that are so ubiquitous (e.g. PubMed/MEDLINE, Entrez Gene, MeSH, the Gene Ontology, and GENIA) that it is difficult to have an extended discussion of BioNLP without being familiar with them. Ethical and legal issues that are specific to the biomedical domain will be covered, including HIPAA regulations (which govern privacy and access to medical records, among other things).

2.1 Early work in the medical domain

The earliest work on biomedical natural language processing was done with inputs from the medical domain. The early work on extraction and synthesis of useful information from text in clinical records and biomedical publications was motivated by increasing concerns over the quality and rising costs of medical care, and the need to keep up with biomedical research. The need for keeping up with a professional field is traditionally satisfied through library services. It is not surprising, therefore, that with the increasing volume and availability of information in the 1960s came an understanding that "the full text of the entire library ought to be read and searched for each request" and that this task could be accomplished only by a computerized system (Swanson 1960). Similarly, computer-based systems were needed to facilitate improvements in medical care services, epidemiological and clinical research, and planning of medical care resources. Parties involved in these services need information extracted from medical records (for the most part kept as free text) and relevant valid findings of medical research extracted from the literature (US Congress 1977).

Subsequently, the observation that some logical relations between entities could be implied through piecing together facts scattered in different sets of publications (Swanson 1986b) brought about text mining in its strict sense of discovery of new facts and hypothesis generation. Text-based knowledge discovery relies on text mining in its broader sense and uses extraction of overtly stated pertinent information to infer relationships between biomedical entities and processes.

The first system capable of detecting medical terms in free text using pattern matching was ELIZA (Weizenbaum 1966). In its DOCTOR mode, the system mimicked a psychiatric interview using canned phrases and key words detected in the sentences entered by a user. ELIZA lacked a knowledge base and real understanding of the submitted text, which, among other factors, such as syntactic and semantic ambiguity, were listed as problems in one of the earliest analyses of medical language processing (Pratt & Pacak 1969).

Focusing on pathology diagnoses, Pratt *et al.* outlined a system for automated processing of medical English (APME), which would take medical text as input and generate "a linguistic description and semantic interpretation of the given utterance." The APME system used the Systematized Nomenclature of Pathology (SNOP) as its lexicon base. SNOP provided terms in four semantic categories, semantic relations between the terms, and a primitive grammar. A morphological analyzer was used to map the terms found in the text to those contained in SNOP. Transformational and paraphrasing rules based on syntactic and semantic features were used to find synonyms (for example, *atrophic muscle* and *muscle atrophy*). Noting that "the boundaries of the semantic unit do not correspond, necessarily, to the boundaries of the noun phrase," the authors introduced the notion of the "kernel phrase" and syntactic and semantic rules for identification of kernel phrases. Once the kernel phrases were identified, the semantic relationships between the phrases were established using a set of relational predicates (Pratt & Pacak 1969).

Other early computer systems employed in healthcare delivery used natural language processing to identify biomedical concepts in patients' problem lists (a common feature of documents in health records) for subsequent data analysis. For example, the SCAMP system was used to examine the incidence and clinical characteristics of skin reactions to co-trimoxazole (Shapiro 1980). The medical language processing systems that leveraged resources developed in the 1960s within the Linguistic String Project (LSP) mapped the facts extracted from clinical text into a structured database, in order to retrieve, compare and summarize the data (Friedman *et al.* 1983).

Presently, many major clinical centers continue developing and using sophisticated systems for extraction of facts and reasoning with clinical narrative. For example, the Medical Language Extraction and Encoding System (MedLEE) uses a lexicon to map terms into semantic classes and a semantic grammar to generate formal representation of sentences (Friedman 2005). MedLEE was used in adverse event discovery (Hripcsak *et al.* 2003), abstraction of findings related to risks of developing breast cancer (Xu *et al.* 2004), and in establishing the significance of associations between diseases and drugs (Chen *et al.* 2008).

2.2 The emergence of the biological domain

The 1990s saw drastic changes in the way that biological scientists worked. While previous assays might get them information on just one gene at a time, and a working biologist might be familiar with just a few genes, new assays made possible by developments in genomics and bioinformatics suddenly made it possible for them to get information on tens of thousands of genes in a single experiment. This might yield a list of hundreds of genes that behaved significantly differently from the other genes in the organism. Understanding these patterns of gene behavior would involve reading about them, but the sheer size of the data sets made this a daunting task. Meanwhile, efforts like GenBank, the earliest genomic database (dating to 1982), were building massive databases of all known genes in multiple organisms. These databases became valuable to the extent that they included information beyond the simple existence of the genes, such as information about gene function. However, populating the databases with this information required reading about them. Simultaneously, the number of biomedical journal publications was rising exponentially (Hunter & Cohen 2006), making keeping up with the publications in one's field a daunting and even impossible task. Furthermore, it increasingly became necessary to read the literature in other fields, besides one's own, as it became clear that many genes had multiple functions and were studied in disparate academic areas. As early as 1994, genomic scientists realized that the field of natural language processing might have something to offer them (Hafner *et al.* 1994).

The earliest papers on actual biological text mining systems appeared in 1998. A set of three papers presented that year and the next year tackled the problems of named entity recognition (Fukuda *et al.* 1998), discussed in detail in Chapter 3; rule-based information extraction (Blaschke *et al.* 1999), discussed in detail in Chapter 3.6; and machine-learning-based information extraction (Craven & Kumlien 1999), discussed in detail in Chapter 3.6. The subsequent years saw an enormous amount of activity in biomedical NLP, much of it by biologists and bioinformaticians, and some of it somewhat naive with respect to what had already been accomplished in the general NLP world.

The early development of biological text mining was driven by biologists motivated by various tasks they had to perform to estimate the scope of existing findings relevant to their own molecular biology research. Early on, creation of repositories of facts about specific genes for species of interest (model organism databases) was driving research in triage of biomedical literature and information extraction. For example, one of the early systems, MedMiner, was developed for literature exploration of gene-drug and gene-gene relationships (Tanabe *et al.* 1999). MedMiner relied on information about genes contained in the GeneCard database and extracted additional information from PubMed/MEDLINE abstracts (the primary database of biomedical journal publications, discussed in detail in Chapter 11). The iHOP system organized initial literature search results around the gene or protein of interest and provided information about all other genes and biomedical terms identified in search results (Hoffmann & Valencia 2004). Textpresso, initially associated with WormBase, extracted sentences from scientific publications and linked gene names identified in the sentences to WormBase entries (Müller, Kenny, & Sternberg 2004). The early literature based discovery experiments performed by hand by Swanson (Swanson 1986a) were later assisted by the ARROWSMITH application (Smalheiser & Swanson 1999).

2.3 Clinical text mining

Mining of clinical text is viewed primarily as a means for improving healthcare outcomes, whereas mining of biological publications aims to advance knowledge of biology and medicine by means of facilitating discovery and understanding of fundamental physiological and pathological processes. Despite the differences in intermediate aims, the end-goal of both sub-domains of biomedical text mining is facilitating improvements in quality of life through understanding of life processes. Methods employed in mining of clinical and biological text are conceptually similar. In both sub-domains, search for facts needed to satisfy specific goals often starts with information retrieval methods applied to collections of text documents. Once the set of documents is reduced to those potentially containing necessary information, complex computationally-intensive natural language processing methods are applied to the smaller set of documents. The borders between the traditional goals of the sub-domains are beginning to disappear in the area of translational research focused on delivery of personalized medical care based on the results of fundamental biological studies. However, echoing the differences in intermediate aims of the two sub-domains, finding and extracting information from biomedical literature for understanding of life functions requires knowledge distinctly different from the clinical domain, as well as understanding of disparate types of text.

Nevertheless, differences in approaches persist. Much of this is due to the differing availabilities of data and of tools in the two sub-domains. The clinical domain has always been plagued by a lack of availability of primary data. In contrast, the biological domain has been supplied with a vast quantity of copyright-free data via PubMed/MEDLINE. Furthermore, the types of data that are relevant to clinical NLP are extremely varied, ranging from agrammatical hand-written notes (Stetson *et al.* 2002) to journal articles. In contrast, the data that is relevant to the biological domain is much more homogenous, almost entirely consisting of edited publications. Tools differ for the two domains as well; for example, workers in the medical domain have long benefited from the existence of lexical resources like the UMLS (described in Chapter 7.1) and conceptual processing tools like MetaMap (presented in Chapter 6.3), while lexical resources have only become available in the biological domain much more recently, and recognizing concepts in the biological domain remains a very open research project.

2.4 Types of users of biomedical NLP systems

A variety of different user types are potential consumers of biomedical natural language processing systems. Much recent research has focused on the needs of model organism database curators. *Model organisms* are organisms that we study because they (like most eukaryotes) share many of the genes of humans, on whom we cannot experiment; and because they are good for studying particular aspects of biology. For example, mice are strongly genetically similar to humans, so they are used as experimental models for many diseases. The worm C. elegans has exactly 969 cells, and we know the exact genesis of each one of them, so they are very well suited for studying cell development and differentiation. The zebrafish D. rario has a transparent body, so it is well-suited for studying phenotypes that involve changes in internal organs. Model organism databases contain information about genes in a specific model organism. None of the information in these databases comes from the knowledge of the individual curator, or database maintainer; rather, every piece of information is linked to a specific publication that provides evidence for it. Thus, tools for working with scientific publications are potentially very useful to model organism database curators. A variety of types of tools are useful to them, including tools for document triaging and named entity recognition tools.

Bench scientists, or biologists doing experimental work, can benefit from text mining as well. In the recent past, a biologist might be intimately familiar with one gene and have a passing acquaintance with a dozen more. Experimental techniques allowed for studying only one gene at a time. Since the mid-1990s, new experimental techniques, known as *high-throughput assays*, have allowed experimentalists to obtain information on as many as 30,000 genes in a single experiment. Hundreds of genes may be found to behave statistically significantly differently from other genes in such an experiment. Most of these genes will be unfamiliar to the researcher. Any gene may have thousands of publications associated with it, and it is impossible for the researcher to master this body of literature in order to gain an understanding of the roles of these many genes in the experimental condition. Text mining can be used to quickly extract relevant information from these many papers, building sets of assertions by information extraction or summarizing publications. Text mining techniques have also been found useful in data analysis algorithms, e.g. by using publications to improve clustering of genes or to filter results.

Within the health care domain, there is similarly a variety of potential user types. Clinicians can benefit from systems for question-answering or for better document retrieval. Text mining has also been found useful for increasing the productivity of *coders*, whose job it is to assign disease and procedure codes to patients for billing purposes.

2.5 Resources and tools

Domain knowledge becomes accessible in a systematic way when its standard representation is stored in publicly available repositories. Access to knowledge is facilitated by tools and meta-data describing the nature of knowledge and the form in which it is stored. The biomedical domain is extremely rich in publicly available resources. One resource lagging behind in this domain (for reasons explained in the next section) is clinical text.

Biomedical domain knowledge resources of a variety of types are available. Some are repositories focused on experimentally determined macromolecular structural data (for example, Protein Data Bank). Others are bibliographic citations into biomedical literature, such as MEDLINE. Additionally, there are tools that help access the repositories, such as the suite of Entrez tools developed at the National Center for Biotechnology Information, US National Library of Medicine (NLM) to access over 30 databases maintained by NLM. Some of the resources and tools widely used in contemporary text mining applications are described below.

US National Library of Medicine

The US National Library of Medicine started out as a resource for military physicians (the Surgeon General's Library) in 1818 (Blake 1986). In the late 1870s, Dr. John Shaw Billings expanded the initial collection that contained about 2,300 medical volumes to a collection of over 120,000 volumes and introduced his system of cataloging the library's collection. Dr. Billings also created the first comprehensive index of journal articles in the Library's collection, Index Medicus. The library continued to grow, overcoming difficulties and changing locations, until the 1956 National Library of Medicine Act, sponsored by Senators Lister Hill and John F. Kennedy, established the institution as the national library in the Public Health Service.

In the 1960s, NLM started developing specifications for a computerized system for producing the monthly Index Medicus. In 1964, NLM inaugurated the Medical Literature Analysis and Retrieval System (MEDLARS), which provided individual bibliographies through its specialized "demand search" service. It soon became apparent that to keep up with the demand for specialized services, NLM needed to research and test the means for communicating biomedical information and evaluate information retrieval techniques, software, and systems. The NLM Board of Regents and staff drafted a policy that stated the need for "a national resource for information systems research and development relevant to human health" and "a clearing-house and coordinating agency for information systems R and D within the Public Health Service." To address these needs, the Lister Hill National Center for Biomedical Communications was established in 1968. In 1988, the National Center for Biotechnology Information (NCBI) was established as a national resource for molecular biology information. The need for specialized resources focused on biomedical research was recognized by Senator Claude Pepper, who sponsored legislation that established NCBI. NCBI was established as a division of NLM due to its experience in creating and maintaining biomedical databases, and because as part of NIH, it could establish an intramural research program in computational molecular biology.

MEDLINE database

MEDLINE (Medical Literature Analysis and Retrieval System Online), a large bibliographic database maintained by NLM, is known as an authoritative and comprehensive source of references into peer reviewed fundamental biomedical studies, as well as clinical evidence. In a 2001 BMJ editorial, it was called one of America's greatest gifts to the world, and the best free starting point for finding high quality medical information (Smith & Chalmers 2001). MEDLINE contains over 19 million references to articles from approximately 4,800 biomedical journals in 30 languages, dating back to the 1950's. With the exception of a few weeks for a yearly update, 1,500 to 3,500 references are added to the database every day. The scope of the database is biomedicine and health, broadly defined to encompass those areas of the life sciences, behavioral sciences, chemical sciences, and bioengineering needed by health professionals and others engaged in basic research and clinical care, public health, health policy development, or related educational activities. MEDLINE also covers life sciences important to biomedical practitioners, researchers, and educators, including aspects of biology, environmental science, marine biology, and plant and animal science, as well as biophysics and chemistry.

Each MEDLINE citation includes basic information such as the title of the article, authors, journal and publication type, date of publication, language, etc. The majority of publications covered in MEDLINE are scholarly journals. Description of the state of MEDLINE by year maintained by the NLM Bibliographic Services Division can be found at http://www.nlm.nih.gov/bsd/. For about 4,500 journals, MEDLINE provides a link to the publisher's Web site to request or view the full article, depending on the publisher's access requirements. Many publishers provide free full text of the article.

Medical Subject Headings

Additional metadata are associated with each MEDLINE citation. Of these, controlled vocabulary terms are assigned by human indexers employed by NLM. The NLM controlled vocabulary thesaurus, Medical Subject Headings (MeSH), contains approximately 23,000 descriptors arranged in a hierarchical structure and more than 151,000 Supplementary Concept Records (additional chemical substance names) within a separate thesaurus. Indexing is performed by approximately 100 indexers with at least a bachelor's degree in life sciences and formal training in indexing provided by NLM. Since mid-2002, the Library has been employing software (MTI – Medical Text Indexer) that automatically suggests MeSH headings using the abstract of the article being indexed and related MEDLINE citations (Aronson *et al.* 2004). The automatically derived recommendations are consulted for about 40% of MEDLINE indexing. Metadata is provided in several formats, of which MEDLINE format and XML are most widely used. Figure 2.1 shows metadata for an abstract in MEDLINE format. To represent different aspects of the topic described by a particular MeSH heading (descriptor), up to three subheadings (qualifiers) may be assigned. In MEDLINE format the subheadings are indicated by the slash notation. An asterisk placed next to a MeSH term indicates that the human indexer interprets the term to be the main focus of the article. More than one MeSH term can be identified as representative of the focus of the article. In the example, *Warts/*therapy* indicates that the article focuses on treatments for warts. The other two starred MeSH headings describe the treatment options: bandages and cryotherapy. The publication type data indicate that the study was a randomized clinical trial. The *Treatment Outcome* heading indicates that the goal of the study was to determine effectiveness of the treatments.

```
PMID 12361440
...
TI   - The efficacy of duct tape vs cryotherapy in the treat-
       ment of verruca vulgaris (the common wart).
...
LA   - eng
PT   - Clinical Trial
PT   - Journal Article
PT   - Randomized Controlled Trial
PL   - United States
JT   - Archives of pediatrics and adolescent medicine
JID  - 9422751
RN   - 0 (Adhesives)
SB   - AIM
SB   - IM
MH   - Adhesives
MH   - Adolescent
MH   - *Bandages
MH   - Child
MH   - Humans
MH   - *Cryotherapy
MH   - Prospective Studies
MH   - Treatment Outcome
MH   - Warts/*therapy
EDAT - 2002/10/04 04:00
SO   - Arch Pediatr Adolesc Med. 2002 Oct; 156(10): 971–4
```

Figure 2.1 MEDLINE citation in MEDLINE format.

PubMed

MEDLINE is freely accessible on the Web through PubMed, the National Library of Medicine's gateway, or through third-party organizations that license MEDLINE from NLM. PubMed[1] is a Boolean search engine that indexes titles, abstracts and metadata separately. These indices allow users to specify which fields or indices should be searched. For example, tagging the search term as follows: *warts[mh]* indicates that only metadata should be searched for the term. By default, searches that

1. http://www.ncbi.nlm.nih.gov/pubmed/

contain automatically identified MeSH terms are expanded using terms indented under the recognized term in the MeSH hierarchy. For example, searching PubMed index for *warts* is augmented with searches for *Condylomata Acuminata* and *Epidermodysplasia Verruciformis*, unless users indicate that no search expansion is desirable. It is worth mentioning that the wide variety of advanced search options makes PubMed a highly competitive search engine when used by an experienced searcher (Hersh *et al.* 1994).

In addition to PubMed, NLM provides Entrez Programming Utilities for batch retrieval of MEDLINE citations with all capabilities of PubMed.

GENIA

The GENIA (GENome Information Acquisition) project is representative of the efforts geared towards enhancing the MEDLINE knowledge base for specific purposes. The broadly stated goal of the GENIA project is automatic extraction of biochemical informtion from journal papers and abstracts by means of language engineering (Collier *et al.* 1999). Over the years, GENIA has developed into a set of tools and annotated documents for development and use of BioNLP methods. The GENIA corpus consists of 1,999 MEDLINE abstracts retrieved using the terms *human*, *blood cells*, and *transcription factors*. The abstracts are annotated with linguistic and semantic information. The semantic annotation includes terms and events contained in the GENIA ontology. The tools include the GENIA tagger for part-of-speech tagging, shallow parsing and named entity recognition; an annotation tool (XConc) that serves as an XML editor, concordancer, and ontology browser; and a tool for protein–protein interactions extraction.

PubMed Central International

Increasingly, full text of scientific publications becomes publicly available for research purposes through digital archives of biomedical literature. These digital archives include the PubMed Central (PMC) archive developed by NLM. PMC archives articles from journals that deposit material in PMC regularly, as well as manuscripts with results of research funded by NIH and the Wellcome Trust. It is a part of the PubMed Central International network, which currently includes the U.S. PubMed Central and UK PubMed Central. This network was recently joined by PMC Canada, which provides access to all publications resulting from the Canadian Institutes of Health Research.

The Journal Archiving and Interchange Tag Suite developed at NLM provides a common format for publishers and archives to exchange journal content. The XML schema modules within the suite define elements for the textual and graphical content of publications. The common representation and the open access model of these archives geared towards development of the global network for exchange of biomedical information should facilitate the discovery process.

2.6 Legal and ethical issues

Mining of biomedical text pursues the goals of improving healthcare and individual patients' status and as such has to address the moral and legal issues of biomedical research. For example, similar to biologists who should ask themselves if the perceived benefits of an experiment replicating smallpox virus will out-weigh the risks, developers of natural language processing resources and systems for clinical decision support should test their systems for unintended consequences and analyze resources for potential risks of disclosing information about individual patients.

The ethics issues are in general not regulated by laws, but disclosure of personally identifiable information often contained in clinical text is strictly regulated in many countries. For example, in the United States, the Health Insurance Portability and Accountability Act of 1996 (HIPAA) Privacy Rule protects the privacy of individually identifiable health information. The HIPAA Privacy Rule applies to health plans, health care clearinghouses, and the electronic health care transactions conducted by providers. In addition to protecting privacy of health information and establishing when such information may be disclosed (including the conditions under which information may be used for research purposes), the Rule gives patients rights over their health information.

It is important to note that Protected Health Information (PHI) excludes health information that is de-identified according to specific standards. De-identified health information can be used in research. There are two ways to determine if health information is individually identifiable: (1) using statistical verification of deidentification (which requires documented certification by a person with appropriate knowledge and experience) or (2) removing 18 elements that could be used to identify the individual or the individual's relatives, employers, or household member from each record and having no actual knowledge that the remaining information could be used alone or in combination with other information to identify the individual who is the subject of the information.

The following identifiers must be removed from the records:

1. Names.
2. All geographic subdivisions smaller than a state.
3. All elements of dates (except year) for dates directly related to an individual.
4. Telephone numbers.
5. Facsimile numbers.
6. Electronic mail addresses.
7. Social Security numbers.
8. Medical record numbers.
9. Health plan beneficiary numbers.
10. Account numbers.
11. Certificate/license numbers.
12. Vehicle identifiers and serial numbers, including license plate numbers.
13. Device identifiers and serial numbers.
14. Web universal resource locators (URLs).
15. Internet protocol (IP) address numbers.
16. Biometric identifiers, including fingerprints and voiceprints.
17. Full-face photographic images and any comparable images.
18. Any other unique identifying number, characteristic, or code, unless otherwise permitted by the Privacy Rule for re-identification.

Understandably, de-identification of substantial text collections is a non-trivial task, which explains the lag in release of clinical text collections (Sibanda & Uzuner 2006). Although there has been considerable progress on the de-identification process per se in recent years (Uzuner, Luo, & Szolovits 2007), the legal and ethical considerations remain.

2.7 Is biomedical natural language processing effective?

A small but growing number of studies have addressed the question of whether or not BioNLP actually makes a contribution to the work of biomedical scientists and clinicians. Overall, the results have encouraging. Some of the relevant studies are described below.

Dowell *et al.* examined the contribution of gene mention recognition (see Chapter 3) to the gene indexing task in a model organism database, and estimated that indexers went from being able to index about 50 documents a week to being able to index 60–70 documents a week (Dowell *et al.* 2009).

The PreBIND system was built to assist in the curation of a database of protein–protein interactions. It uses a statistical system to find relevant documents, and then a rule-based system to locate protein names. The BIND database evaluated it and found that it reduced the duration of one task by 70%, corresponding to a time savings of 176 person days (Donaldson *et al.* 2003).

The Immune Epitope Database built a classifier to determine if documents should be curated or not, classifying them as clearly relevant, clearly irrelevant, or unsure. They reported that this resulted in a speed-up of curation time due to a speed-up in the abstract selection process, but did not measure the exact amount of increase (Wang *et al.* 2007).

The Comparative Toxicogenomics Database built a system for ranking documents retrieved by PubMed/MEDLINE queries and found that there was a strong correlation between the ranking of the documents and whether or not the documents contained curatable data (Wiegers *et al.* 2009).

Systems have also been built that have resulted in the construction of databases based purely on text mining results. These include a database of mutations in G protein-coupled receptors and nuclear hormone receptors (Horn, Lau, & Cohen 2004) and post-translational modifications (Shah *et al.* 2005). The RLIMS-P program has also been very useful in building the curated Protein Information Resource, extracting substrates, enzymes, and phosphorylation sites (Hu *et al.* 2005; Narayanaswamy, Ravikumar, & Shanker 2005; Yuan *et al.* 2006).

Alex *et al.* examined the effectiveness of an NLP system in aiding curators of protein–protein interactions, and found that the use of text mining tools could speed up curation time by 1/3 if NLP output was completely accurate. However, they also cautioned that time was not sufficient as the only metric to gauge an NLP system's effectiveness (Alex *et al.* 2008).

The nature of the task mostly determines the evaluation metrics. For example, near 100% recall is needed in retrieving the initial set of relevant documents for systematic clinical reviews (Bekhuis & Demner-Fushman 2010). Clinicians looking for evidence to support their decisions need 'bottom-line advice' extracted from highly reliable sources with high precision (Ely *et al.* 2005).

An open question is how to evaluate the contribution of the biomedical NLP processing to its end-goals: advancement in knowledge acquisition and improvements in health-care.

CHAPTER 3

Named entity recognition

3.1 Overview

Named entity recognition (NER) is the task of locating instances of mentions of some specific semantic class or classes in text. NER was the earliest topic tackled in the modern genomic era of BioNLP. It has continued to be an active topic of research to the present day. One way in which NER differs in the BioNLP area is the diversity of semantic types of named entities that must be handled. Early Message Understanding Conference shared tasks introduced the familiar categories of PERSON, ORGANIZATION, and LOCATION (Sundheim 1992). NER applications for crime and intelligence analysis have added the categories of. However, the size of the set of potentially important named entities in biomedicine dwarfs these lists. Table 3.1 lists just some of the semantic categories of named entities that have already been tackled in the biomedical domain. These barely scratch the surface of the types of entities that would be useful for fully general text mining purposes.

Table 3.1 A sample of the semantic classes of named entities that must be recognized in biomedical NLP. Note the surface similarities between many of the examples. Adapted from Jurafsky and Martin (2008).

Semantic class	Examples	Systems
Cell lines	*T98G, HeLa cell, Chinese hamster ovary cells, CHO cells*	Settles (2005); Bada & Hunter (2007)
Cell types	*primary T lymphocytes, natural killer cells, NK cells*	Settles (2005); Johnson *et al.* (2006); Bada & Hunter (2007)
Chemicals	*citric acid, 1,2-diiodopentane, C*	Johnson *et al.* (2006); Corbett, Batchelor, & Teufel (2007)
Drugs	*cyclosporin A, CDDP*	Rindflesch *et al.* (2000)
Genes/proteins	*white, HSP60, protein kinase C, L23A*	Yeh *et al.* (2005)
Malignancies	*carcinoma, breast neoplasms*	Jin *et al.* (2006)
Disorders	*amyotrophic lateral sclerosis*	Aronson (2001a)
Mouse strains	*LAFT, AKR*	Caporaso *et al.* (2005)
Mutations	*C10T, Ala64 → Gly*	Caporaso *et al.* (2007)
Populations	*judo group*	Demner-Fushman & Lin (2007)

3.2 The crucial role of named entity recognition in BioNLP tasks

NER plays not just a crucial but a primary role in many BioNLP tasks. The paradox of much of BioNLP is that NER is assisted by a number of preprocessing tasks – tokenization and parsing, for example – but these preprocessing tasks are made easier if NER is done first. Take the task of tokenization. As we have seen in an earlier chapter, tokenization in biomedical text poses unique problems. For chemicals, the tokenization problems that we have seen earlier are all obviated if chemical named entities are recognized and marked as tokens themselves before tokenization is attempted. Consider the impact of gene names on syntactic parsing. Recognizing gene names can be facilitated by locating the boundaries of syntactic phrases. However, consider the impact of gene names like *upregulated during skeletal muscle growth 5* (Entrez Gene ID 66477, symbol Usmg5). A syntactic parser attempting to analyze this gene name would have to solve a part-of-speech ambiguity for the word *upregulated*, determine a prepositional phrase attachment, and decide what to do with the terminal numeral. However, if we can precede parsing by labelling this named entity as a named entity, we reduce some of the burden on the parser and avoid a number of sources of error for a stretch of text that actually does not need to be parsed at all.

Similarly, in the case of corpus construction, one of the points of the construction effort is to locate named entities, but a number of tasks can be avoided – part-of-speech tagging, treebanking, and predicate/argument marking – if we can locate the named entities before corpus construction begins. Of course, the remainder of the text will need to be POS-tagged, etc.

3.3 Why gene names are the way they are

Historically, gene names were the first focus of genomics NER, and they have continued to be an active target of research. (Genes are related to proteins and other "gene products" – at the risk of a great deal of simplification, there is a one-to-one relationship between genes and proteins. See Hunter (2009) for a text that is accessible to natural language processing specialists that lays out the true complexities of the situation.) For NER purposes, gene and protein names are almost always considered to be completely equivalent, although genes and proteins are very different things[2].

2. See (Hatzivassiloglou, Duboué, & Rzhetsky 2001) for a rare attempt to differentiate between them that shows how difficult the task is, even for humans.

One thing that makes gene NER difficult is that there is a tremendous diversity in the possible forms of gene names. For one thing, a typical gene has both a "name" and a "symbol," and from an NER perspective, both are considered "names." In the case of the example given earlier in this chapter, *upregulated during skeletal muscle growth 5* is the gene's name, and *Usmg5* is its official symbol. As we will see in the chapter on concept normalization, the term "official" is something of a misnomer; authors are in no way constrained to use the "official" names or symbols in their writing, and they may not even know what they are.

The diversity of types of gene mentions in text goes far beyond the differences between "names" and symbols, though. To understand this diversity, it is helpful to know some of the history of the naming of genes.

Just as the discoverer of a new species is given the privilege of naming it, the discoverer of a new gene is given the privilege of naming it. The first gene to be discovered was a fruitfly gene which, when mutated, causes a fly to be born with eyes that are white, rather than their normal red color. This gene was named *white* (symbol *w*, Entrez Gene ID 31271), beginning a tradition of naming genes after the physical appearance that a fly takes on when they are mutated. Thus, we have genes named *swiss cheese, clown, daschund*, and *dreadlocks*.

When it was found to be possible to study the genetics of behavior, genes began to be named not on the basis of physical appearance, but on the basis of how flies act when they are mutated. Thus, we have genes named *ether a go-go, lush, agnostic*, and *amontillado*.

The next step was metaphorical gene names. For example, the *lot* gene was named because experimenters initially thought that flies with mutations in this gene were averse to salt; in the Bible, Lot's wife turns to salt. Embryonic development stops in *maggie* mutants; similarly; the Simpsons character Maggie never ages from infancy. One pole of the mitotic spindle does not migrate to the end of the cell in flies with a mutation in *scott of the antarctic*; Robert F. Scott was an explorer who failed to reach the South Pole.

Almost no gene or protein acts alone. Rather, genes interact in complicated ways, often in networks. As genes that interacted with other genes were discovered, they were sometimes given names that were related to the name of the gene with which they interacted. Thus, there is a gene named *sevenless*, so called because when it is mutated, flies are born without the seventh facet of their faceted eyes; the following genes were given their names based on their interactions with the sevenless gene:

- *bride of sevenless* (symbol *boss*, Entrez Gene ID 43146)
- *son of sevenless* (symbol *sos*, Entrez Gene ID 34790)
- *daughter of sevenless* (symbol *dos*, Entrez Gene ID 38321)

Some of these expressions of interrelatedness were expressed metaphorically. For example, a gene that was found to interact with the *always early* gene was named *british rail*. A gene that was found to interact with the *asp* gene was named *cleopatra*.

Eventually, whimsy became acceptable in the fruitfly community. Thus, one lab named every gene that it discovered after a different alcoholic beverage, and another has given sixteen genes Slavic female names.

Obviously, many of these gene names are common English words. Some of them contain stopwords. More insidiously, a number of gene symbols are isomorphic with function words, such as *a* (full name *arc*, Entrez Gene ID 43852), *A* (full name *abnormal abdomen*, Entrez Gene ID 43851), *It* (full name *irregular teeth* (not a fly gene), Entrez Gene ID 103984), and *And* (full name *androcam*, Entrez Gene ID 44913).

This does not exhaust the variety of naming strategies for genes. In the human and mouse research communities, it is common to name genes with a description of their function, e.g. *GM2 ganglioside activator* (Entrez Gene ID 2760); with the name of a disease with which they are associated, e.g. *multiple sclerosis* (symbol *MD*, Entrez Gene ID 126200), *breast cancer 3* (symbol *BRCA3*, Entrez Gene ID 605365), *cystic fibrosis intestinal distress* (symbol *Cfid*, Entrez Gene ID 100036221); or with some other phrase that describes some aspect of their behavior, such as the previously described *Usmg5*, *inracisternal A particle-promoted polypeptide* (symbol *ipp*, Entrez Gene ID 448063), and *BMP and activin membrane-bound inhibitor* (symbol *bambi*, Entrez Gene ID 493556).

Thus, the range of possible forms of gene names is enormous, ranging from very short symbols to multi-word names containing or consisting of function words, punctuation, and common English words. This diversity of form poses a formidable challenge to computational systems.

One might assume that gene names could be identified in text by using one of the databases of genes (for example, Entrez Gene, GeneCards or MGI) as a gazetteer. This turns out to work poorly, for a variety of reasons. One is that the putatively "official" names in these gazetteers are by no means treated as official by the community, and they are often used in non-canonical forms (e.g. the name *hemoglobin, beta* Entrez Gene ID 3043 might appear as *beta-hemoglobin* (PMID 18728021)) or not used at all – this may vary even within a single document. A second reason is that such a gazetteer would contain numerous common English words, for the reasons described above, causing numerous false positives (see Morgan *et al.* 2004) for an extensive analysis of this phenomenon). Finally, new genes are constantly discovered, and new names are coined both for new genes and previously known ones. It *is* possible to deal with the first and last of these problems in a very sophisticated gazetteer-based approach (Hanisch *et al.* 2005) that does considerable preprocessing of the data, but results from the BioCreative gene NER task show that in the majority of cases, incorporating a gazetteer into a gene NER system is not helpful (discussed further below).

3.4 An example of a rule-based gene NER system: KeX/PROPER

One of the first papers of the modern genomic era in BioNLP was a paper on a named entity recognition system (Fukuda *et al.* 1998). For some years, it was the most heavily cited paper in BioNLP (393 citations as of June 2009, per Google Scholar). This system focussed on gene names, particularly in yeast. It is a classic example of a rule-based system.

The authors distinguish between two elements of gene names: core terms and feature terms. Their *core terms* are distinguishable by containing sentence-medial capital letters, numbers, and non-alphanumeric characters, e.g. *Src, p53,* or *Brca-1.* Their *feature terms* are a fixed set of keywords describing the function or nature of a core term, e.g. *receptor* or *protein,* as in *p53 protein.*

The method of the system is to first locate core terms, and then extend the boundaries of names by recognizing feature terms that go with them. Core term recognition is a five-stage process. In the first stage, candidate core terms are recognized, and in the subsequent four steps, likely false positives are eliminated. The first step – identification of candidate core terms – labels any sentence-medial mixed case tokens, numbers, or non-alphanumeric symbols. For example, this step would identify *Src, site-specific,* +/-, and *99%.* (Of all of these, only *Src* is a legitimate name.)

The first elimination rule targets any token whose length is greater than nine characters and that consists only of lower-case letters and hyphens. This eliminates tokens like *site-specific*, but allows the retention of actual gene symbols like *PPAR-g* (PMID 15665586). The second elimination rule targets any token of which greater than 50% of its characters are non-alphanumeric. This eliminates tokens like +/-, but allows the retention of actual symbols like *PPAR-γ* (PMID 16720732). The third elimination rule targets tokens and substrings that represent units, such as *microM* and %. Finally, the fourth elimination step looks for tokens that are part of certain templates, such as author names in citations, eliminating token sequences like *W. J. Wilbur*. In combination, these four steps would eliminate all of our example candidates from the preceding paragraph except for *Src,* the one legitimate name.

Having found core terms, they are then connected with appropriate feature terms. Two sets of rules are used for this, one involving simple adjacency and the other involving POS tags. The first rule simply connects together any core terms or feature terms that are adjacent to each other. For example, given the core term

Src, followed by the core term *SH3*, followed by the feature term *domain*, these would be concatenated to yield the single entity *Src SH3 domain*. (Note again that this system targets subsequences of genes and proteins; if it targeted only genes and proteins, then *Src* would be the sole entity in this example.)

The next set of rules makes use of POS tags. First, it looks for core or feature terms that are separated only by nouns, adjectives, or numerals. For example, given the sequence

Ras	guanine	nucleotide	exchange	factor	Sos
core	NN	NN	NN	feature	core

...the system returns the single entity name *Ras guanine nucleotide exchange factor Sos*.

A second POS-dependent rule extends the name boundary leftward from a core or feature term up to a determiner. For example, given

the	focal	adhesion	kinase	(FAK)
DET			feature	core

...the system extends the boundary leftward from *kinase* to *focal*, yielding *focal adhesion kinase (FAK)*.

A final rule extends the boundary rightward if a core or feature term is followed by a single uppercase letter or a Greek letter. E.g., the core term *p85* followed by *alpha* is combined to form the single entity name *p85 alpha*.

Two final clean-up rules reduce false positives[3]. One looks for any feature words that have not been attached to some core words. This prevents isolated feature words like *kinase* from being returned as gene names. The second rule moves a rightmost boundary leftward if the rightmost element is not a noun, so that *Src-related* would be trimmed to *Src*.

The KeX/PROPER system illustrates what would become a pattern in the construction of many later systems: any techniques to help adjust the boundaries of a named entity tend to improve system performance. Base noun phrase shallow parsing is one example. A popular one is the abbreviation definition technique of Schwartz & Hearst (2003). It has been used to find the full extent of the span to the left of a parenthesized gene symbol, such as *breast cancer associated 1 (BRCA1)*. This is a common pattern for the first appearance of a typical gene symbol in a text.

One of the findings of the first BioCreative shared task was the demonstration of the long-suspected fact that gazetteers are typically of little use in GM.

3. An additional set of rules addressing conjunctions in names is omitted since it is not relevant to most formulations of the GM problem.

BioCreative I allowed participants to submit output in either or both of two categories, called *open* and *closed*. Closed-category output was produced entirely by the automated system, without recourse to outside resources. Open-category output allowed the use of outside resources, such as gene name dictionaries. Figure 3.1 (reprinted from Yeh *et al.* 2005) shows each system's output, with the open and closed results grouped. As can be seen, using a gazetteer was only helpful in the case of a system whose "closed" performance was abysmal, i.e. the systems at the far right of the graph. If gazetteers are used carefully and are subjected to considerable preprocessing, then it *is* possible to make good use of them in GM – see Morgan *et al.* (2004) for the specific case of fly names, and Hanisch *et al.* (2005) for the more general case. However, raw gene name dictionaries have generally not yielded good results; this finding has been replicated in the case of the TREC Genomics shared tasks.

The BioCreative GM corpora remain the standard data sets for evaluating GM systems.

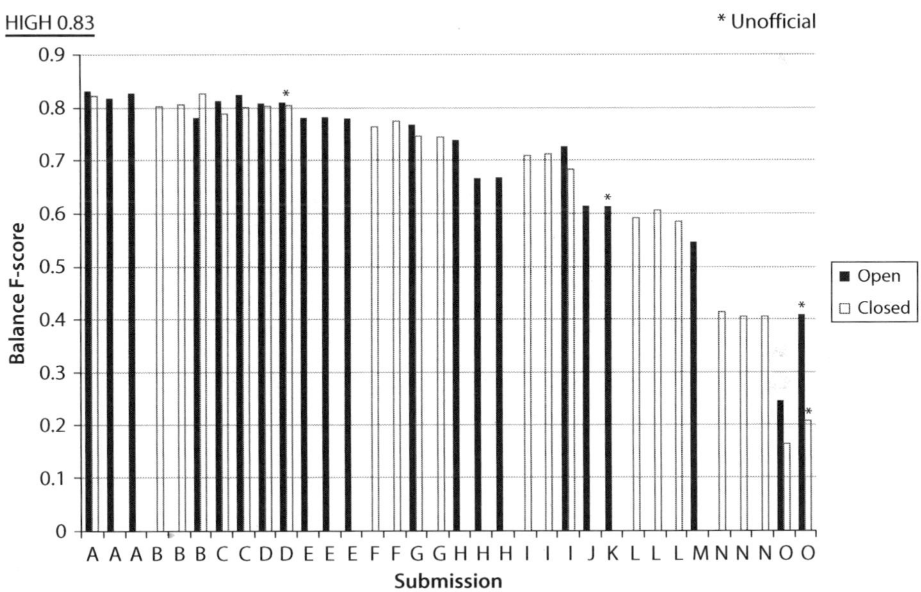

Figure 3.1 Balanced F-scores of the submissions to the BioCreAtIvE Task 1A (gene mention finding evaluation).

3.5 An example of a statistical disease NER system

Historically, many medical NER systems are dictionary-based due to richness of domain knowledge resources such as the UMLS Metathesaurus. (The Unified Medical Language System is described in Chapter 11.) However, even the most comprehensive knowledge sources do not fully cover all known named entities and, above all, terminology for new diseases, conditions, drugs, and other rapidly evolving areas of biomedical knowledge. For example, at the time of the 2003 SARS (Severe Acute Respiratory Syndrome) outbreak, the only sense of SARS in the UMLS Metathesaurus was *SARS gene* (seryl-tRNA synthetase).

Additional senses and concepts can be captured using corpus-based methods, such as the weakly-supervised approach to recognition of diseases in MEDLINE abstracts of randomized clinical trials based on iterative pattern learning (Xu *et al.* 2008). In this approach, recognition of disease names is bootstrapped using several seed patterns pertaining to diseases and commonly found in publications. New patterns are learned from the extracted diseases, and new disease names are discovered iteratively. The process is stopped when it reaches a pre-defined number of iterations. This approach was inspired by bootstrapping methods for finding instance-concept relations (Hearst 1992). Approaches based on iterative automatic acquisition of surface patterns associated with named entities could potentially learn spurious patterns and erroneously label instances. For example, in *skills needed to manage patients with complex problems*, "complex problems" could be labeled as a specific disease name. Pattern and instance ranking methods could potentially improve the accuracy of pattern selection and disease extraction.

The bootstrapping algorithm proposed in Xu *et al.* (2008) starts with a seed pattern – for example, *patients with NP*, where NP is a disease name such as *hepatitis C virus infection* found immediately to the right of the seed pattern in the sentence parse tree. In the first iteration over MEDLINE abstracts all NPs to the right of the seed patterns are extracted as disease names and added to the list of disease name instances. Then, for each name on the disease name list, bigrams (two tokens) to the left of each disease name instance are extracted as additional patterns. For convenient retrieval, the sentences extracted from the abstracts of randomized clinical trials and parsed using the Stanford Parser, were indexed using the publicly available search engine API Lucene. This retrieval system allows finding all pattern and disease name instances as needed.

The extracted patterns are ranked compared to the seed patterns using the disease name instances found by both types of patterns. The three explored ranking methods favor:

- *sensitivity*: the ratio of diseases found by both the original seed and the new pattern to the number of diseases found by the seed pattern alone. This score favors general patterns and is prone to noise.
- *specificity*: the ratio of diseases found by both the original seed and the new pattern to the the number of diseases found by the new pattern alone. This score favors patterns associated with very few diseases.
- *balance*: the harmonic mean of the first two scores will penalize patterns associated with few diseases.

Information about pattern rank is subsequently used in ranking the candiadate disease name instances. The disease name ranking methods are based on:

- *abundance* (document frequency) – the number of documents containing a disease name;
- *number of patterns* that recognized the disease name; and
- *best patterns* – the number of times the disease is associated with the highest ranking pattern among those that recognized the disease.

The evaluation of the algorithm on 100 manually annotated abstracts demonstrated that the best pattern disease ranking method combined with the cutoff level at top 5% of the disease list improves disease name recognition, compared to dictionary-based methods. The improvements in F1 score (0.81, compared to 0.60) are achieved through higher recall (0.78 at 0.80 precision).

This method has several limitations: it will propagate parser errors in identifying noun phrase boundaries. The patterns generally vary in length, whereas the method limits the size to two words. Disease names and NP boundaries do not always coincide. However, the method is fairly robust with respect to the seed patterns: different seed patterns result in finding the same top-ranking new patterns.

3.6 Evaluation

Named entity recognition systems are evaluated using the usual metrics of precision, recall, and F-measure.

The BioCreative definitions of the GM problem in BioCreative I and II have become the standard definitions of the task. However, others have existed, particularly in earlier publications. Some of the dimensions on which these definitions vary include:

- What exactly are the definitions of the semantic classes that are being targetted? For example, the KeX/PROPER work included domain names, motifs, sites, and fragments. Some work has included RNA molecules, while other work has excluded them.
- What should be included within the boundaries of a gene name? For example, given an input like *human BRCA1 gene*, different authors might define the intended right answer as *human BRCA1 gene, human BRCA1, BRCA1 gene*, or *BRCA1*. The BioCreative organizers attempted to deal with this by listing multiple principled correct answers in their gold standard.
- Should all boundaries be treated equally in scoring? See Olsson *et al.* (2002) on the score effects of different boundary decisions.

Until the 2007 community-wide evaluation that involved assigning ICD-9 billing codes (a standardized vocabulary for representing disorders and treatments) to pediatric radiology reports (Pestian *et al.* 2007), there were only individial research efforts in evaluating clinical named entity recognition. Since then, the i2b2 NLP challenges were largely focused on NER in clinical text. Similarly to the BioCrative evaluations, the i2b2 organizers are exploring nuances of NER annotation boundaries, annotator consensus and importance of entity types and attributes. The first i2b2 shared task (2007) involved recognition of all entities needed for de-identification of clinical data in accordance with HIPAA rules. Other challenges involved extraction of medications and other attributes of a drug prescription, and extraction of disorders and diagnostic and therapeutic procedures.

Information about collections and tools developed for the biomedical domain can be found in the Online Registry of Biomedical Informatics Tools (ORBIT)[4]. ORBIT is a community-wide effort to create and maintain a structured, searchable metadata registry for informatics software, knowledge bases, data sets, and design resources, facilitated jointly by the Veterans Administration and NLM.

4. http://orbit.nlm.nih.gov/

CHAPTER 4

Relation extraction

4.1 Introduction

Information extraction (also known as relation extraction) is generally defined as the extraction of information about a very focussed type of relationship from free text. Historically, it can be traced back to the Message Understanding Conferences (MUC). The MUC shared tasks defined information extraction in terms of templates that defined specific relations between semantically defined types of entities. These relations identified specific types of events (using that term in its nontechnical sense) that can be found in newswire articles, such as terrorist attacks and corporate succession. Each relationship type was represented by a frame – an information structure that bundles together all of the participants in an action, represented as slots in the template. The requirements of the task, then, were to recognize that an event had occurred and to fill the slots with the appropriate participants. The classic MUC system is a rule-based system called FASTUS (Hobbs *et al.*), described in detail in Jackson & Moulinier (2002).

Much work on information extraction in the biomedical domain has represented a sort of step backwards from the MUC tasks in terms of ambition, in that the frames that are targetted by biomedical information extraction systems are typically only binary in arity – that is, they target relations between just two entities (Blaschke *et al.* 1999; Craven & Kumlien 1999; Ono *et al.* 2001; Blaschke & Valencia 2001; Bunescu *et al.* 2005; Huang *et al.* 2004b; Krallinger, Leitner, & Valencia 2007; Rosario & Hearst 2005; Krallinger *et al.* 2008) (among many others). Historically, one of the most common information extraction targets has been that of protein–protein interactions (PPI).

4.1.1 Protein–protein interactions as an information extraction target

Almost no protein does its job in isolation. Rather, proteins act as members of networks of proteins in which they interact with one or more other proteins, either directly or indirectly, e.g. by regulating the activity of other proteins or by executing non-consecutive steps in a series of direct interactions that carry out some biological process.

Protein–protein interactions may take place between pairs of proteins, for example when one protein transfers a molecule to another, or may take place between groups of proteins, either directly, by forming a "complex" or assembly of different proteins, or indirectly, as for example being members of the same network of multiple proteins. Figure 4.1 (Varadan *et al.* 2005) shows both a pair of proteins and a protein complex. Understanding protein–protein interactions is one of the crucial tasks in understanding both normal physiological function on the one hand and many human diseases on the other (Kann *et al.* 2006). Building databases of protein–protein interactions has been the subject of major efforts, including the Database of Interacting Proteins (Xenarios *et al.* 2002), MINT (Chatr-aryamontri *et al.* 2006), and IntAct (Kerrien *et al.* 2006). Due to their importance, many biomedical information extraction systems have targeted the extraction of information about protein–protein interactions.

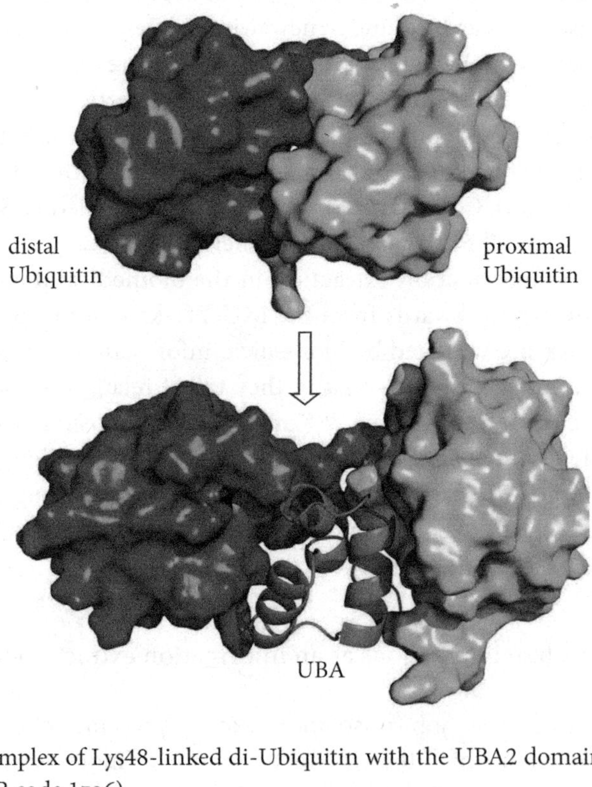

Figure 4.1 Complex of Lys48-linked di-Ubiquitin with the UBA2 domain from hHR23A. (PDB code 1zo6)

4.2 Binarity of most biomedical information extraction systems

Although as described above, many protein–protein interactions take place between groups of proteins, as when forming a complex, the tendency has been to represent protein–protein interactions as binary from the point of view of information extraction templates. One shared task on the subject (Krallinger *et al.* 2008) even required systems that recognized complexes to break down those complexes into n^2 binary interactions.

Information extraction systems have been built for other binary relationships, including gene/disease, protein/subcellular location, and protein/function. However, with few exceptions, these have tended to be binary, as well, and have tended to target only a single relationship type. Moving beyond this has been one way that researchers have attempted to move the field forward.

4.3 Beyond simple binary relations

Most biomedical information extraction systems have targetted just a single type of biomedical event. However, a number of systems have attempted to move beyond this by recognizing multiple subclasses of a single event type.

An early example of a supervised machine learning approach to recognizing different events involving the same entity types comes from the work of Rosario & Hearst (2004). They built a system that broke down the single broad relation of treatment/disease into eight categories of relations, such as:

- CURE: *Intravenous immune globulin for recurrent spontaneous abortion*
- PREVENT: *Statins for prevention of stroke*
- CAUSE: *Malignant mesodermal mixed tumor of the uterus following irradiation*

This work is especially significant from a knowledge representation point of view because the participants are the same for all relation types, making simple approaches such as co-occurrence inapplicable, and even weakening very knowledge-intensive approaches. The authors annotated a corpus of sentences from PubMed/MEDLINE abstracts and tested a variety of machine learning algorithms and features for differentiating between these relations, including words, part of speech, shallow parses, and crucially, the semantic feature of MeSH ID for words for which these could be found. Some orthographic features were used, as well. They achieved accuracy of around 80% when semantic roles were not given and 97% when they were.

In later work, they extended this concept of fine-grained relation identification to differentiate between ten different types of protein–protein interaction (Rosario & Hearst 2005). Following an independent third-party classification scheme from the HIV-1 protein interaction database, they tackled ten relation types:

- *degrade*
- *synergizes with*
- *stimulates*
- *binds*
- *inactives*
- *interacts with*
- *requires*
- *upregulates*
- *inhibits*
- *suppresses*

One of these (*interacts with*) is something of a catch-all category, and some of these are overly lexically differentiated and could reasonably be collapsed (*inhibits* and *suppresses*), but the fact remains that the set of categories represents a far more nuanced approach to protein–protein interactions than had been seen in any previous (or most subsequent) work on the problem. Their learning-based system achieved an accuracy of 64% for this 10-way distinction.

Chun *et al.* (2006) returned to a more clinically oriented task. They began by annotating a set of 3,939 sentences that contained both a gene mention and a term related to prostate cancer with respect to thirteen distinct topics:

- study description (method)
- modality
- genetic variation
- epigenetics
- gene expression
- gene produces variation
- molecular function
- sub-cellular localization
- pharmacology
- clinical marker
- risk factor
- tumor biology
- remarks

They then looked at inter-annotator agreement for these thirteen categories between four annotators and found that IAA was high enough to justify using only six of these. So, they ended up with the following six categories:

- study description (method)
- genetic variation
- gene expression
- epigenetics
- pharmacology
- clinical marker

They then trained a machine learning based system to differentiate between these six categories, with overall results of 92.1% precision over all relations. One of the findings of this study was that different feature sets yielded the best results for different categories. For example, for study description and genetic variation, the most contributory feature was bag of words, and for gene expression, epigenetics, pharmacology, and clinical marker, the order of candidate features was the most contributory feature. (Despite these differences, the overall sets of features for the various relation types, although they had some differences, were quite similar.) One notable feature of the work was that specific genes contributed to performance, raising the question of how well a similar system would scale to new discoveries.

The BioNLP'09 shared task (Kim *et al.* 2009) went in a somewhat different direction. It required participants to recognize eight different event types, plus semantic modifications of these events, such as negation and speculation. The number of participants in the events varied, with some types of events having only a single participant and some optionally having a second participant. The eight event types were:

- *gene expression* (single participant)
- *protein catabolism* (single participant)
- *localization* (single participant required, originating and ending locations optional)
- *binding* (multiple binding participants required, binding sites optional)
- *phosphorylation* (single participant required, phosphorylation site optional)
- *regulation* (one participants required, an optional Cause argument can be a protein or another event)
- *positive regulation*
- *negative regulation*

Like Rosario and Hearst's work and the other papers cited in this section, this work was important from a knowledge representation perspective because multiple relations shared semantic types of participants (all eight relations involved proteins), making simple co-occurrence or even knowledge-based semantic-constraint-based systems inapplicable. The work was also innovative in that the regulation events could have another event as their cause.

The preceding systems have all made the notion of binary relations more complicated, but remained binary in terms of the arity of the templates that they assumed. A small number of systems have tried to tackle greater-than-binary relations.

The RLIMS-P program (Hu *et al.* 2005; Narayanaswamy, Ravikumar, & Shanker 2005; Yuan *et al.* 2006), available on-line, targets a three-way relationship between protein kinases, protein substrates of phosphorylation, and phosphorylation sites. It is also interesting in that it uses a small set of rules to find slot fillers across sentence boundaries.

Lu (2007) describes a system that targets a four-way template for protein transport. The four slots are the transported protein, the transporting protein, the starting location of the protein, and the ending location of the protein. This system was unusual in that it employed an ontology both to define the information extraction task, using an adaptation of a portion of the Gene Ontology related to protein transport, and also to constrain the semantic participants in transportation events. Unlike most of the systems that we have discussed so far, which were machine learning based, this was a rule-based system. It is discussed in further detail in Hunter *et al.* (2008).

GeneWays (Rzhetsky *et al.* 2004) extracts information about molecular pathways from full-text articles, then aggregates and visualizes the results. GeneWays consists of four modules:

1. A tagger that identifies the following entities: complex, disease, domain, gene, gene or protein, process, protein, species, and small molecule.
2. An NLP GENomics Information Extraction System (GENIES, based on MedLEE, described later in this chapter) that identifies nested relationships between the extracted entities and outputs machine-readable trees.
3. The Simplifier module converts the nested relationships to binary statements that link two entities through an action, for example, interleukin-2 binds interleukin-2 receptor. The binary statements are stored in the Knowledge module.
4. The Visualization module graphically presents the pathways in CUtenet figures.

The number of displayed relations is regulated by a document frequency threshold (for example, displaying relations occurring in at least 10 papers). GeneWays search is available at http://anya.igsb.anl.gov/genewaysApp/actionsearch.htm.

Evidence suggests that greater-than-binary representations are, in fact, appropriate for the biomedical domain (Wattarujeekrit, Shah, & Collier 2004; Kogan *et al.* 2005; Cohen & Hunter 2006). From a knowledge representation perspective, one promising move in that direction has been the beginnings of development of predicate-argument-structure-based shallow semantic representations for biomedical verbs, in the style of PropBank (Palmer, Gildea, & Kingsbury 2005) and VerbNet (Kipper-Schuler 2005). However, the bridge from verbal representations to template frames has not been broadly attempted.

Assuming such a bridge, PAS representations lead naturally to approaching biomedical information extraction as a semantic role labelling task. Bethard *et al.* (2008) used the data from Lu (2007) to train and evaluate a semantic role labelling system. Bethard *et al.* found a number of differences between typical newswire-based semantic role labelling tasks and this biomedical task, many of them related to aspects of the input text. They noted that nominal predicates and light verbs predominated, unlike in typical Wall Street Journal text. They also noted that lexical content was very different from typical newswire text, with many predicates present that are not found at all in PropBank (see also Wattarujeekrit *et al.* in this respect). They also noted that unlike typical semantic role labelling systems, which take syntactic units in a parse tree as primitives, they had to use smaller chunks to accommodate for the fact that many arguments of nominalizations were smaller than traditional syntactic units, particularly in the case of compound noun phrases like *fatty acid transport protein translocation*, where *fatty acid transport protein* is an embedded transported object. Overall, they were able to achieve precision of 87.6% and recall of 79.0% when using manually annotated protein boundaries, showing the upper boundary of performance if GM were perfect, and precision of 87.0% and recall of 74.5% for the more realistic case when protein boundaries were determined automatically.

4.4 Rule-based systems

Many early approaches to information extraction in the biomedical domain are characterizable as rule-based systems. Some of them were based on FASTUS-like approaches, but other early ones are notable for the fact that they were built using little or no linguistic information. These systems are of interest to the computational linguist because they provide very good baseline systems, showing the limiting case of what can be accomplished with no linguistic information while still sometimes exhibiting good performance.

4.4.1 Co-occurrence

Some of the earliest attempts to build protein–protein interaction information extraction systems with extremely high coverage were based on the concept of co-occurrence: that is, two proteins co-occurring in a textual unit of some defined size. Jenssen *et al.* (2001) looked for co-occurrences of gene and protein names in 10,000,000 PubMed/MEDLINE abstracts and used such co-occurrence to posit a network of 13,712 human genes. Edges between genes were weighted by the number of articles that contained the pair. Coverage was assessed by comparing the set of edges to the pairs of interacting genes in the Database of Interacting Proteins, and was found to include 51% of the pairs in DIP. Precision was not assessed, but the method was found to give insight into biological processes reflected in large gene expression experiments.

More recent work has focussed on fine-tuning the method used to determine whether or not to use a particular instance of co-occurrence in constructing a network. Research by Gabow *et al.* (2008) is representative. They sought to calculate a confidence measure for a given interaction predicted by co-occurrence in text. However, unlike previous work, they wanted to calculate an asymmetric measure, favoring proteins that are under-represented in text. The rationale behind this is useful from a biological research perspective, since understudied proteins are of particular research interest. The formula that they derived is the Asymmetric Co-occurrence Fraction (ACF):

$$ACF\,(x, y) = \frac{n_{\mathrm{xy}}}{\min[n_{\mathrm{x}}, n_{\mathrm{y}}]}$$

... where n_x is the number of documents that mention gene x, n_y is the number of documents that mention gene y, and n_{xy} is the number of documents that mention both x and y.

The Asymmetric Co-occurrence Fraction was evaluated extrinsically on the task of protein function prediction. It was compared to two biological data sources – genetic interaction and protein–protein interaction – and to two other metrics for assessing co-occurrence in literature. The evaluation was done for three separate organisms – yeast, worm, and fly (see Figure 4.2). The Asymmetric Co-occurrence Fraction was found to make a positive and in some cases striking improvement to protein function prediction. This was especially notable in the case of the worm, since there is relatively little protein–protein interaction data available for this organism as compared to the others.

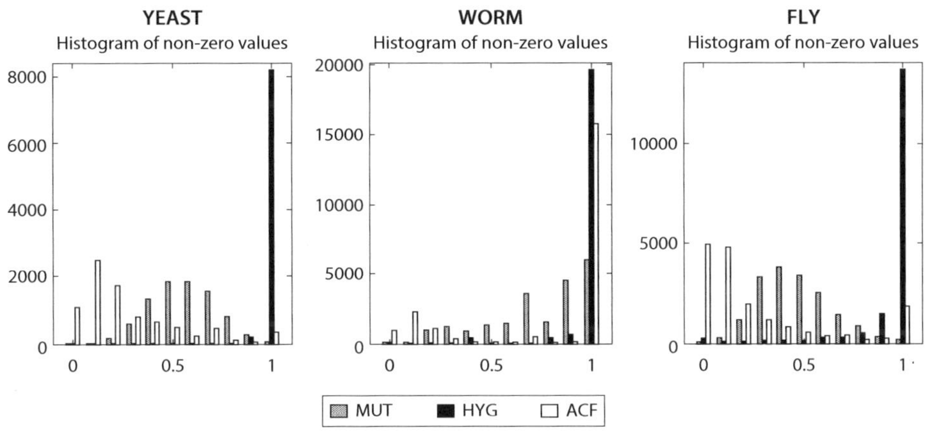

Figure 4.2 Histogram Comparison of Co-Occurrence Measures. Histogram of the number of proteins assigned a given confidence value by the co-occurrence measures. Abbreviations: MUT – Mutual Information Measure; HYG – Hypergeometric Measure; ACF – Asymmetric Co-occurrence Fraction.

4.4.2 Example rule-based systems

The two earliest papers on genomic information extraction were both published in 1999. One of them took a rule-based approach, while the other took a machine learning approach (Craven & Kumlien 1999). We describe the rule-based system here.

In a trend that we will see repeatedly in this book, the system was built not by language processing specialists but by a group of highly motivated biologists.

The system begins with an information retrieval step in which queries are constructed containing particular genes of interest. The system then splits abstracts into sentences and further subdivides the sentences into smaller chunks demarcated by the punctuation marks .,; (period, comma, and semicolon). Within these chunks, it performs the simple task of looking for any two proteins separated by an "action word." The system used a very small set of such action words, consisting of the following lemmata:

- acetylate (-ed, -s, -ion)
- activate (-ed, -s, -ion)
- associated with
- bind (-ing, -s, -s to, bound)
- destabilize (-ed, -s, -ation)

- inhibit (-ed, -s, -ion)
- interact (-ed, -ing, -s, -ion)
- is conjugated to
- modulate (-ed, -s, -ion)
- phosphorylate (-ed, -s, -ion)
- regulate (-ed, -s, -ion)
- stabilize (-ed, -s, -ation)
- suppress (-ed, -es, -ion)
- target

The evaluation was unusual. Rather than calculating precision and recall on their data set, the authors evaluated the system on its ability to reconstruct two known interaction networks. For example, the system attempted to reconstruct the system known as Pelle in the fruit fly. This system involves eight interactions between the six proteins *spatzle, toll, pelle, tube, cactus,* and *dorsal.*

The tool correctly reconstructed eight of eight edges in the interaction graph between these six proteins, with only one false positive edge. The results presented in Blaschke *et al.* (1999) are shown in Figure 4.3.

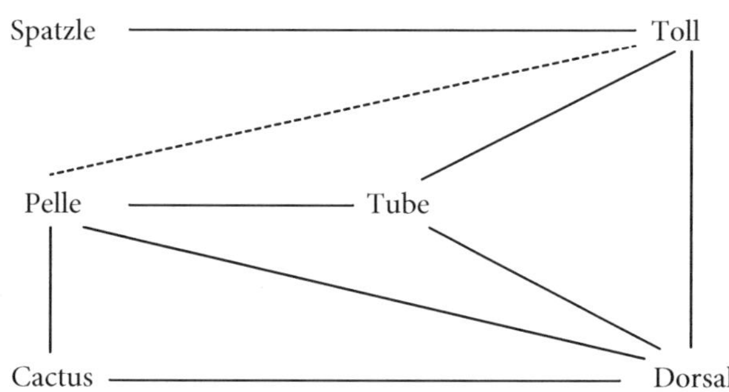

Figure 4.3 Relations between proteins in the Drosophila Pelle system.

Two things about this system should be noted: it succeeded in uncovering a biologically important set of facts – something that has not even been attempted in more recent systems, which typically target only isolated facts using the familiar P/R/F-measure evaluation – and it did so using no linguistic information at all.

This was an impressive accomplishment, and shows that similar systems can serve as a non-trivial baseline even for modern systems.

A later version of the system, known as SUISEKI, introduced four new features into the system (Blaschke & Valencia 2001).

- The restricted list of verbs was replaced with a set of patterns that allowed for any verb to be used in a pattern.
- Part of speech tagging was introduced to allow for finding the above-mentioned verbs.
- Patterns that handled negation were used.
- Distance between proteins and verbs was added as a feature.

Different patterns could then be assigned different estimates of correctness. Extracted facts could then be merged or excluded depending on whether or not they met a threshold. An example of the new type of rule in this version of the system is

> [protein] (0–5 words) [verb] (0–5 words) [protein] probability score = 4
> [protein] (6–10 words) [verb] (6–10 words) [protein] probability score = 2

Rule-based systems continue to be used to the present day. One finding from these systems is that at least for constrained tasks, rule-based systems can be built with far less effort and far fewer rules than has previously been assumed. For example, only five rules were required for the protein transport extraction task described elsewhere in this chapter; they achieved an F-measure of 0.59. One of the highest performances in the BioNLP '09 shared task on event extraction (see above) was achieved by Kilicoglu & Bergler (2009), who wrote a total of only 27 rules for extracting events and event participants from dependency trees.

Systems like Kilicoglu and Bergler's use sophisticated linguistic information. However, high-performing information extraction systems have continued to be built in the spirit of Blaschke *et al.*'s original work, using relatively small sets of highly lexicalized manually generated patterns without linguistic information. For example, the OpenDMAP system placed first in the BioCreative II protein–protein interaction task with a system that used no linguistic information and a set of only nine rules (Baumgartner Jr. *et al.* 2008). OpenDMAP (Hunter *et al.* 2008) is a system that makes extensive use of simple lexicalized patterns in conjunction with a relatively large amount of domain knowledge.

4.4.3 Machine learning systems

The first system to apply machine learning to a problem in the genomics domain is described in Craven & Kumlien (1999). They focussed on learning the subcellular localization of proteins. Proteins are targetted towards particular locations within the cell, and knowledge of the subcellular localization of a protein is important both in its own right and because knowing the subcellular localization of a protein can help us make predictions about its likely function (Ng 2006).

The named entity recognition component of the system was simple; the system targetted only the six proteins serotonin, secretin, NMDA receptor, collagen, trypsinogen, and calcium channel. The system relied on exact matching to recognize mentions of these proteins, which was probably safe in the case of most of the six.

The feature representation was a simple bag-of-stems model. They then built a simple Bayesian classifier and trained it on all of the individual sentences in a set of 2,889 MEDLINE abstracts using Laplace smoothing. Simple co-occurrence of proteins and subcellular compartments was used as a baseline. Their system outperformed the baseline at a wide range of values for recall (see Figure 4.4).

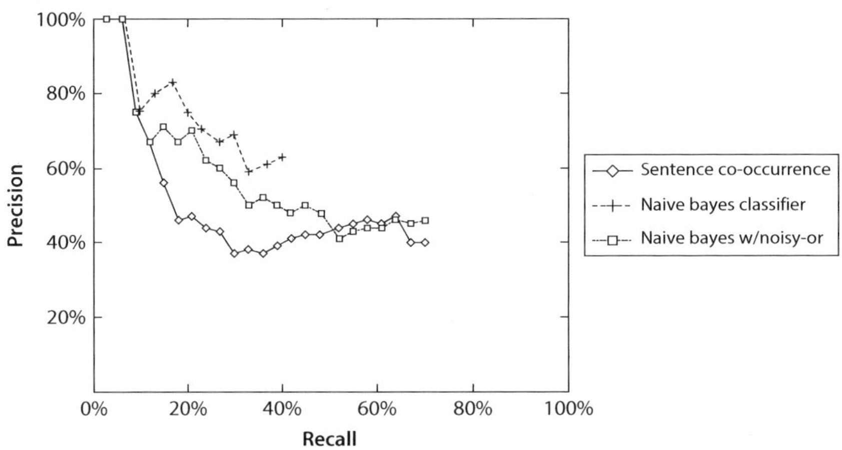

Figure 4.4 Precision versus recall of the subcellular localization extraction system in Craven & Kumlien (1999).

Their paper lists the words that were found to be highly weighted for predicting positive instances. It includes words that are obvious indicators of subcellular localization, such as *local, insid, immunofluoresc, immunoloc, accumul,* and *microscopi*. However, it also includes words that are not obviously related to subcellular localization. The authors attribute this to the difficulty of reliably estimating the probability of a word belonging to the positive or the negative class based on limited training data.

Subsequent applications of machine learning to the task of information extraction in the biomedical domain have used much more sophisticated feature sets (Craven and Kumlien also used parse trees in further experiments reported in the same paper), and have applied the more rigorous approach of applying actual gene mention recognition to arbitrary proteins. However, this paper provided an early proof-of-concept that machine learning techniques are applicable in the biomedical domain and served as the foundation for a considerable amount of subsequent work.

Table 4.1 Log of the ratio of the probability of a word *w* occurring in a positive document to the probability of that word occurring in a negative document for the top twenty values of the words in a vocabulary of 2,500 stems (Craven & Kumlien 1999).

local	.00571
pmr	.00306
dpap	.00259
insid	.00209
indirect	.00191
galactosidas	.00190
immunofluoresc	.00182
secretion	.00181
mcm	.00157
mannosidas	.00157
sla	.00156
gdpase	.00156
bafilomycin	.00154
marker	.00141
presequ	.00125
immunoloc	.00125
snc	.00121
stain	.00115
accumul	.00114
microscopi	.00112

4.5 Relations in clinical narrative

We distinguish clinically relevant relations, such as *drug* TREATS *disease*, extracted from published literature and those extracted from clinical documents (patients' discharge summaries, progress notes, radiology reports, etc.). The need for this distinction is illustrated by modifications necessary to adapt MedLEE (Friedman 2005), a system developed and widely used for processing clinical text, to process biomedical publications (Chen & Friedman 2004). In this section we will focus on processing of clinical narrative and present MedLEE, currently the most sophisticated NLP system integrated with a Clinical Information System. One of the distinguishing features of clinical narrative is that in many cases it needs to describe events that are not present, occurred in the past or might occur in the future, or happened to patient's relatives. For example, a discharge summary usually presents a patient's medical history, family history, and instructions given to the patient on discharge. It is common to see phrases like "a long standing history of peripheral vascular disease", "patient's coworker had a cold", and "in case of worsening cough, call your family doctor". Negation is a particularly common modification of clinical events. For example, if a patient had a chest x-ray "to rule out pneumonia", the radiology report might state that there was "no evidence of pneumonia". After discussing MedLEE, we will present NegEX (Chapman *et al.* 2001), a system specifically developed to handle negation and certainty in clinical narrative.

4.5.1 MedLEE

The Medical Extraction and Encoding (MedLEE) system, originally developed at the New York-Presbyterian Hospital to process radiology reports (Friedman *et al.* 1994), interprets clinical narrative of cardiology and pathology reports, discharge summaries, etc. using a grammar that integrates syntactic and semantic components.

```
<Sentence>      -->  <Patterns> ("."|";") .
<Patterns>      -->  <FindingRel> {<MoreFinding>} .
<FindingRel>    -->  <FindingPhr> | <BpMods> <Verbrel> <FindingPhr> .
<FindingPhr>    -->  {<Lmods>) <Findingterm> {<Rmods>) .
<MoreFinding>   -->  <Relation> <FindingRel> .
<Findingterm>   -->  disease | cfinding | pfinding } descriptor .
<Lmods>         -->  [<CertMods>] [<DegreeMods>] [<ChangeMods>] [<BpMods>] .
<Rmods>         -->  <SpatialRel> <BpMods> | <Lmods> .
<CertMods>      -->  [negation] certainty | negation .
<Relation>      -->  conjunction | <Verbrel> .
<Verbrel>       -->  [auxverb] [be] [negation] certainty .
<DegreeMods>    -->  degree .
<ChangeMods>    -->  [negation] change  .
<BpMods>        -->  {<RegionMods>} bodyloc {<MoreBpMods>) .
<MoreBpMods>    -->  <SpatialRel> <BpMods> | conjunction <BpMods> .
<SpatialRel>    -->  in | on | at | along | near | under .
<RegionMods>    -->  region {<MoreRegion>) .
<MoreRegion>    -->  conjunction <RegionMods> .
```

Simplified semantic grammar for radiology findings <...> = name of a non-atomic semantic structure defined in the grammar, "..'' = literal, [...] = optional element, {...} = symbol may occur zero or more times. (Re-printed from Friedman 2005 with the author's permission.)

To handle various types of reports, MedLEE needs a preprocessor for each report type. The preprocessor has to establish document sections, add missing sentence endings, and expand abbreviations. For example, given a sign out note without a header, the preprocessor will add a section header, SUMMARY. Abbreviation expansion is lexicon based and domain specific. For example, to correctly interpret the abbreviation *P.E.* as *physical examination* or as *pleural effusion*, the preprocessor needs a correctly set domain variable.

The MedLEE core NLP processor consists of three modules:

- The parser that uses a semantic grammar and a semantic lexicon (which classifies words and multi-word phrases and specifies their canonical forms). The grammar includes translation rules (omitted in the example) that identify semantic components of a given structure, interpret relations between the components, and translate the grammatical structures into pre-specified target forms. The target output forms are a composition of the output structures of each of the identified components. A complete parse of a given sentence is not always achievable due to out-of-vocabulary terms, or ungrammatical structure of the sentence. In these cases, the parsing strategy is relaxed to first ignore unknown words, and then segment the sentence and parse the segments.
- The phrase regularizer that regularizes the output forms of phrases that are not contiguous using a knowledge base that contains compositional structures of decomposable phrases. For example, the phrases "heart is enlarged", "enlarged heart", "heart shows enlargement" and "cardiac enlargement" are normalized to "enlarged heart".
- An optional encoder module that uses a table of codes to translate the regularized forms into unique concepts corresponding to a clinical controlled vocabulary.

For each report, MedLEE produces a set of primary findings and associated modifiers. A sample MedLEE output for two primary findings (one problem and one procedure) and the sentence submitted to MedLEE are shown in Figure 4.5.

4.6 SemRep

SemRep is a rule-based symbolic system that extracts semantic relationships (predications) from biomedical literature. SemRep starts processing with underspecified (shallow) syntactic analysis based on the SPECIALIST Lexicon and the MedPost part-of-speech tagger. The noun phrases identified in the first step are mapped to UMLS concepts by using MetaMap. Semantic relationships, for

```
<sentence>
    <structured form = “xml”>
        <problem v = “vaso occlusive crisis” code = “UMLS:C0750151_vaso occlusive crisis” idref = “p22”>
            <certainty v = “high certainty idref = “p16”></certainty>
            <parsemode v = “model”></parsemode>
            <quantity v = “multiple” idref = “p20”></quantity>
            <sectname v = “report past medical history item”></sectname>
            <sid idref = “s1”></sid>
            <status v = “need” idref = “p28”></status>
            <status v = “past history” idref = “p8”></status>
            <timeper v = “admission” idref = “p32”>
                <quantity v = “multiple”idref = “p30”></quantity>
                <service v = “hospital” idref = “p38”>
                    <location v = “to” idref = “p34”></location>
                </service>
            </timeper>
            <code v = “UMLS:C0750151_vaso occlusive crisis” idref = “p22”></code>
        </problem>
        <problem v = “aplastic crisis” code = “UMLS:C0302111_aplastic crisis” idref = “p41”>
            <parsemode v = “model”></parsemode¿
            <sectname v = “report past medical history item”></sectname>
            <sid idref = “s1”></sid>
            <code v = “UMLS:C0302111_aplastic crisis” idref = “p41”></code>
        </problem>
        <problem v = “urinary tract infection” code = “UMLS:C0042029_urinary tract infection” idref = “p46”>
                <certainty v = “high certainty” idref = “p44”></certainty>
            <parsemode v = “model”></parsemode>
            <sectname v = “report past medical history item”></sectname>
            <sid idref = “s1”></sid>
            <code v = “UMLS:C0042029_urinary tract infection” idref = “p46”></code>
            </problem>
        <procedure v = “transfusion” code = “UMLS:C1879316_transfusion (procedure” idref = “p55”>
            <certainty v = “high certainty” idref = “p51”></certainty>
            <parsemode v = “model”></parsemode>
            <quantity v = “many” idref = “p53”></quantity>
            <sectname v = “report past medical history item”></sectname>
            <sid idref = “s1”></sid>
            <code v = “UMLS:C1879316_transfusion (procedure)” idref = “p55”></code>
        </procedure>
    </structured>
    <tt>
        <sent id = “s1”>The patient’s
        <phr id = “p8”>past medical history</phr>
        is
        <phr id = “p16”>significant for</phr>
        <phr id = “p20”>multiple</phr>
        <phr id = “p22”>vaso-occlusive crisis</phr>
        <phr id = “p28”>requiring</phr>
        <phr id = “p30”>multiple</phr>
        <phr id = “p32”>admissions</phr>
        <phr id = “p34”>to</phr>
        the
        <phr id = “p38”>hospital</phr>
        <phr id = “p39”>, </phr>
        <phr id = “p41”>aplastic crisis</phr>
        <phr id = “p44”>, </phr>
        <phr id = “p46”>urinary tract infection</phr>
        <phr id = “p51”>, </phr>
        <phr id = “p53”>many</phr>
        ¡phr id = “p55”>transfusions</phr>
        .
        </sent>
    </tt>
</sentence>
```

Figure 4.5 Sample MedLEE XML output, courtesy Carol Friedman.

example, "cryotherapy TREATS verruca vulgaris" are identified through syntactic and structural phenomena called indicators. Constraints on allowed relationships are encoded in over 200 manually created indicator rules that map syntactic elements (such as verbs and nominalizations) to predicates in the Semantic Network, such as TREATS, CAUSES, and LOCATION OF. The indicator rules take into account coordination, relativization, and negation.

Dependency grammar rules that enforce syntactic constraints are used to identify arguments of a semantic relationship. For example, the proposition *cryotherapy TREATS verruca vulgaris* is derived from the phrase: *cryotherapy in the treatment of verruca vulgaris* because the indicator **in** points to the Semantic Network relationship *Therapeutic or Preventive Procedure – treats – Disease or Syndrome*, and the arguments *cryotherapy* and *verruca vulgaris* are mapped to concepts having the allowed semantic types. SemRep also handles semantic interpretation of comparative structures restricted to the semantic group Chemicals and Drugs. A set of rules identifies two types of comparison relations: the first asserts only that two drugs are compared; the second provides additional information about the scale on which the drugs are compared, for example, effectiveness, and the relative position of the drugs on the scale, for example, lower than.

SemRep was adapted to identify semantic predications on the genetic etiology of disease. This modification (called SemGen) is now folded into Enhanced SemRep. The main consideration in creating SemGen was the identification of gene and protein names as well as related genomic phenomena. SemGen relies on AB-Gene, in addition to MetaMap and the Metathesaurus. Since the UMLS Semantic Network does not cover molecular genetics, ontological semantic relations for this domain were created for SemGen. The allowable relations were defined in two classes: gene-disease interactions (ASSOCIATED WITH, PREDISPOSE, and CAUSE) and gene-gene interactions (INHIBIT, STIMULATE, and INTERACTS WITH).

The Enhanced SemRep was subsequently expanded to identify predications relevant to pharmacogenomics and not identified by either SemRep or SemGen (Ahlers *et al.* 2007). The UMLS Semantic Network used by SemRep was modified in order to accommodate semantic relations crucial to pharmacogenomics as follows:

Genetic Etiology:
Substance ASSOCIATED WITH | PREDISPOSES | CAUSES *Pathology*
Substance Relations:
Substance INTERACTS WITH | INHIBITS | STIMULATES *Substance*
Pharmacological Effects:
Substance AFFECTS | DISRUPTS | AUGMENTS *Anatomy* | *Process*

Clinical Actions:
Substance ADMINISTERED TO *Living Being*
Process MANIFESTATION OF *Process*
Substance TREATS Living Being | Pathology
Organism Characteristics:
Anatomy | *Living Being* LOCATION OF *Substance*
Anatomy PART OF *Anatomy* | *Living Being*
Process PROCESS OF *Living Being*
Co-existence:
Substance CO-EXISTS WITH *Substance*
Process CO-EXISTS WITH *Process*

4.6.1 NegEX

NegEx takes a sentence with indexed findings phrases as input and determines if a given finding is negated. "Finding" is defined as a UMLS concept in the following semantic types: Finding, Disease or Syndrome, Sign or Symptom, Congenital abnormality, Acquired abnormality, Lab result, Injury or Poisoning, Biologic function, Physiologic function (e.g., energy expenditure, fetal development, Mental process, Mental or Behavioral dysfunction, Cell or Molecular dysfunction, Anatomic abnormality, or Experimental model of disease).

Whereas the original algorithm negated all findings to the right of a negation indicator in the sentence, subsequent modifications determine the scope of the negation. NegEX is a stand-alone knowledge-based system that relies on preprocessing of the text by a NER tool. Given a sentence with identified findings, (for example, "pneumonia" could be replaced with its unique concept identifier in the UMLS or another controlled vocabulary in the phrase: "no evidence of pneumonia"), NegEX uses regular expressions and three types of phrases stored in its knowledge base to identify negations. NegEX skips over the first type of phrases, called "pseudonegation", because these are not reliable indicators of negation (for example, "not ruled out"). The second type of phrases, true negation indicators, are divided into Pre-UMLS negation phrases (phrases that occur before the term they are negating) and Post-UMLS negation phrases (phrases that occur after the term they are negating). The Pre- and Post-negation phrases are used in corresponding regular expressions:

- <pre-negation phrase> * <indexed term>
- <indexed term> * <post-negation phrase>

The negation scope (maximally allowed window size between the terms and negation phrases, indicated by asterisks) is normally the end of the sentence. The negation scope can be decreased, if another negation phrase or a conjunction (from the NegEX conjunction list) is found within the window. The scope terminates at the next negation phrase or conjunction.

4.7 Evaluation

In the clinical domain, relation extraction was undertaken within the i2b2 2010 relation extraction task (Uzuner *et al.* 2011). The participants were provided with 394 training reports and 477 test reports (primarily discharge summaries from three hospitals). The training set was annotated with concepts of the types: medical problems, treatments, and tests, as well as relations between the concepts. The relations between *treatments* and *problems* were defined as: *improves*, *worsens*, *causes*, *administered* for, and *not administered*. The relations between problems and tests were *reveals* and *conducted*. Finally, problems could be in an *indicates* relation.

Starting in 2007, the i2b2 data becomes available under data use agreements to the research community at large in a year after each evaluation.

The AIMED[5] and LLL[6] collections were used to evaluate extraction of relations between proteins and genes.

5. ftp://ftp.cs.utexas.edu/pub/mooney/bio-data/

6. http://genome.jouy.inra.fr/texte/LLLchallenge/

CHAPTER 5

Information retrieval/document classification

The information retrieval task is to take a user information need and return a set of documents that satisfy that information need. Typically there is a query construction step involved.

5.1 Background

5.1.1 Growth in the biomedical literature

One of the striking scientific phenomena of the period between the late 1980s and the current time has been a stable growth in the amount of publications in the biomedical literature (see Figure 5.1). This growth in publications has made it difficult, if not impossible, for researchers to keep up with publications in their field. Furthermore, as genomic science has grown, interdisciplinary boundaries have broken down. For example, the gene relaxin has recently been found to play a role in human responses to drug treatment for heart failure (Du *et al.* 2010). This has required cardiologists to research relaxin. However, it turns out that relaxin has been studied for years by obstetricians, since it is also the hormone ripening protein, responsible for preparing the uterus for labor. Thus, cardiologists find that they must not only stay current with their own literature, but also must become familiar with a body of literature from the obstretrics community.

The growth in scientific publications also affects scientists doing high-throughput experiments. Current high-throughput experiments may result in finding a list of hundreds of genes that behave significantly different from other genes in the sample. Researching these genes in the face of such a large body of literature is a daunting challenge.

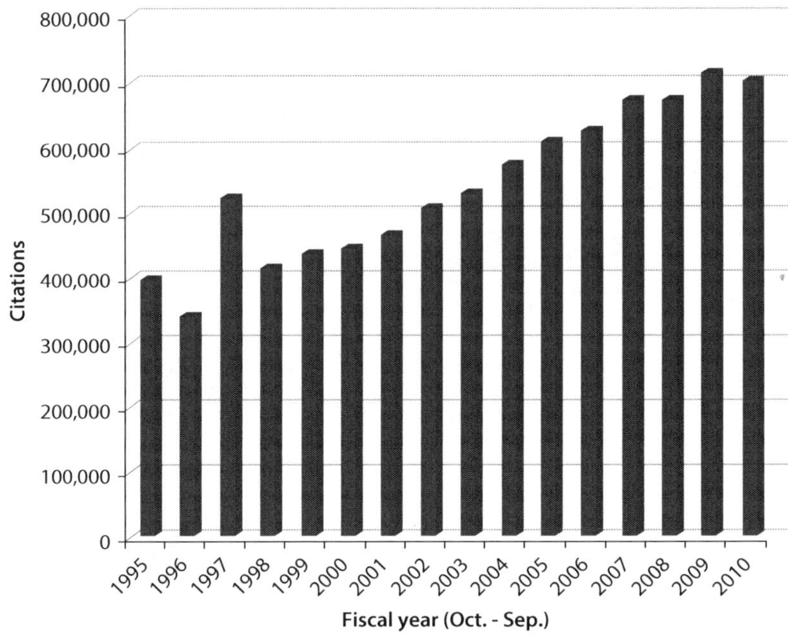

Figure 5.1 The number of indexed citations added to MEDLINE during each fiscal year since 1995. This number does not include OLDMEDLINE subset citations.

5.1.2 PubMed/MEDLINE

MEDLINE is a database of citations (article titles and abstracts and other bibliographic information) in biomedical publications. MEDLINE, which constitutes the major repository of biomedical documents in the world, is most commonly accessed via the PubMed search interface.

The original MEDLINE document collection per se goes back to 1966, when it began with a collection of over 175,000 documents. PubMed experienced a growth spike in 2003 when an over 1.5 million document collection known as OLDMEDLINE was added to it (OLDMEDLINE 2011). The PubMed search engine was instituted in 1996 and currently continues to experience extensive usage, with currently two to three million searches per day. See *A day in the life of PubMed* (Herskovic *et al.* 2007) for more interesting details. PubMed contains many useful features. They change fairly constantly, making any description likely to be out of date by the time of publication, but some interesting additions to the search interface with likely staying power have included:

- The ability to search by MeSH terms
- Suggested spelling corrections
- Listing of possibly related articles
- Links to full-text articles
- Recognition of gene names in queries
- Connectivity with other NLM databases

Despite the fact that PubMed/MEDLINE's usage statistics suggest that it is clearly meeting a need in the biomedical research community, it has come under occasional criticism by the text mining community. The most common criticism is that by default it returns hits in order of publication date. While there is a justification for this – there is an assumption that researchers are most interested in the most recent publications in their area of interest – NLP researchers have often pointed out that it would be helpful to be able to get relevance-ranked results from PubMed/MEDLINE. The Relemed system (Siadaty, Shu, & Knaus 2007) takes an alternative to traditional tf*idf-based systems, scoring articles by the proximity of terms in a multiple-term query. (Most PubMed/MEDLINE queries are claimed to be multiple-term.) The FABLE system, discussed below, also includes an optional relevance-ranking feature.

Another criticism of PubMed/MEDLINE has been that it is limited to searching abstracts, rather than full articles. We will see some of the consequences of this for information retrieval below.

Searches for genes are a perennial problem, compounded by the fact that a researcher may only be interested in the behavior of genes in a specific organism, which may not be reflected in the abstract. The FABLE system (fable.chop.edu), for Fast Automated Biomedical Literature Extraction, is an information retrieval system that was specifically built to allow searching by human gene names. The methodology for gene normalization (see Chapter 6) is described in Fang *et al.* (2006). FABLE also includes an optional relevance-ranking feature, where a variety of features are used to do ranking, including traditional features like term frequency as well as position of a search term in the article and closeness of the match between the actual query term and any synonyms of the query term that might have been found in the article.

5.2 Issues

As in any information retrieval task, synonymy and polysemy pose issues. Including synonyms in queries is known to be helpful for retrieving short texts, although not long ones, and PubMed/MEDLINE abstracts are short texts. However, what is to be used as a source of synonyms, and are there aspects of this domain that make them less useful? In the chapters on lexical resources (Chapter 2 and Chapter 7.1) we talk about the UMLS Metathesaurus and about Entrez Gene and its lists of gene names and synonyms. Briefly, it has been found that such sources of biomedical synonyms are only useful if they are used very carefully (Ando, Dredze, & Zhang 2006; Aronson *et al.* 2005b; Hersh & Voorhees 2008).

For example, the UMLS Metathesaurus, the major component of the UMLS, is constructed from over 100 biomedical vocabularies. Terms from different vocabularies meaning the same thing are grouped together into concepts, and each concept is assigned one or more categories, or semantic types, from the UMLS Semantic Network. This organization of biomedical concepts consisting of surface forms from UMLS constituent vocabularies serves as a powerful basis for supporting biomedical applications, as shown by many studies (Johnson 1999; Friedman *et al.* 2001). However, Metathesaurus content is known to have a number of problems such as missing biomedical concepts, concepts that are not biomedical at all, NOS terms (see Chapter 6 for definition), and ambiguity, perhaps the most important problem with Metathesaurus content.

One source of Metathesaurus ambiguity arises when a concept contains a term which is a substring of the preferred name of the concept but whose meaning differs from that of the concept. For example, the concept *Other location of complaint* contains the term *Other*, which is a spurious synonym of *Other location of complaint*. Note that the Metathesaurus editors have marked some problematic terms as suppressible, making them easy to ignore. Unfortunately, the term *Other* in the above example is not so marked. A source of true lexical ambiguity arises from the existence of acronym/abbreviation terms. For example, the term PAP occurs in fifteen concepts including *Papaverine*, *PAPOLA gene* and *PULMONARY ALVEOLAR PROTEINOSIS, ACQUIRED*. As a final example of true ambiguity – polysemy, consider the term *resistance*, which occurs as a term in the three concepts *Resistance (Psychotherapeutic)*, *resistance mechanism* and *social resistance*. Each of these concepts can legitimately be represented by the homonym *resistance*. The problem in this case is that at least one more legitimate sense of *resistance*, namely *Electrical resistance*, is missing, which is relevant in biomedicine, e.g. in cardiology.

As the Metathesaurus has grown, the goal of effectively using its knowledge has become more challenging, partly due to the growth in ambiguity described above. A large body of work on disambiguation of Metathesaurus homonyms in context provides a means for selecting the correct concept (Humphrey *et al.* 2006; Schuemie, Kors, & Mons 2005; Stevenson *et al.* 2008; Jimeno-Yepes & Aronson 2010). Careful manual clean-up of the synonym sets before use is doable on a per-case basis as well, but may be difficult to carry out automatically on a wide-scale basis. Similarly, Entrez Gene synonym lists may contain essentially meaningless synonyms like *tyrosine kinase* that trigger many hits, may not be weighted low by inverse document frequency, and are not useful gene-specific for information retrieval (ren Fang *et al.* 2006).

One of the first retrieval systems that implemented automatic concept-based indexing and extraction of the UMLS concepts from users' requests was SAPHIRE (Hersh & Greenes 1990). SAPHIRE utilized the UMLS Metathesaurus by breaking free text into words and mapping them into UMLS terms and concepts. Documents were indexed with concepts, queries were mapped to concepts, and standard term frequency and inverse document frequency weighting was applied. When measured with combined recall and precision, SAPHIRE searches performed as well as physicians using MEDLINE, but not as well as experienced librarians (Hersh *et al.* 1994).

Other experiments with synonymy have produced mixed results. Voorhees (1999) found a 13.6% decrease in average precision comparing effectiveness of conceptual indexing with baseline indexing of single words for 30 queries and 1033 medical documents. Srinivasan (1996) demonstrated an overall improvement of 16.4% in average precision, primarily due to controlled vocabulary feedback (expanding queries by adding controlled vocabulary terms). Aronson & Rindflesch (1997) achieved 14% improvement in average precision through query expansion using automatically identified controlled vocabulary terms that were expanded using inflectional variants (gender, tense, number, or person) from the SPECIALIST lexicon (Browne *et al.* 2003) and synonyms encoded in UMLS. Finally, concept-based indexing of MEDLINE citations based on manual and semi-automatic indexing (Aronson *et al.* 2004) is utilized in PubMed.

5.3 A knowledge-based system that disambiguates gene names

We discuss here work on a knowledge-based system that is capable of disambiguating gene names in information retrieval. The work is especially significant because of its demonstration of the feasibility of a clever strategy for evaluating genomic information retrieval. (The TREC Genomics track had tried the technique earlier, but there was not sufficient material available at the time to make it feasible.)

Retrieval of documents when the query is a gene is a perennial problem. The problem is made worse by the extensive ambiguity in gene names that we discuss in the section on gene normalization.

Another problem, this one in the context of experimentation, is how to evaluate biomedical information retrieval systems. TREC-like manual paradigms have been applied, but these are very labor-intensive and difficult to repeat through the course of a day of development. This work addressed that issue, as well.

One well-studied approach to information retrieval in general is query expansion. One possible use of query expansion is to increase recall, although as discussed above, this has potential pitfalls. However, another use of query expansion is for disambiguation purposes. To give a non-medical example, suppose that one wanted to find information on the UNIX command cat. Simply searching on the term *cat* is not likely to be successful, since it will return many documents about felines that are not relevant to this particular search. However, adding other UNIX terms, e.g. *cat more less*, results in UNIX-relevant hits being returned at the top of the list. (Of course, adding the word UNIX would likely have the same effect.) Sehgal and Srinivasan (2006) experimented with a variety of sources of this sort of external knowledge for query expansion. They had in common that they are all publicly available at no cost, being provided as part of the Entrez Gene (LocusLink, at the time) database.

Sehgal and Srinivasan restricted their data set to publications that were associated with GeneRIFs. A GeneRIF, or *Gene Reference Into Function,* is a less-than-255-character segment of text that describes some aspect of gene function, broadly construed. GeneRIFs are a feature of the Entrez Gene database, where they are associated with specific genes. GeneRIFs are generally constructed by cutting-and-pasting them with at most modest modifications from the abstract of a journal article, and each GeneRIF therefore has a PMID associated with it. By restricting their document collection to documents that were associated with a GeneRIF, they assured that they had a gene/document mapping for each document. Any document associated with a gene was relevant to a query constructed for that gene, and any other document from the collection was not relevant. At the time, this yielded a document set of size 9,390[7].

7. Similar methods could also be used to build document sets for other types of queries; for example, the set of documents linked to OMIM entries could probably be used to build a document collection for information retrieval work in Mendelian diseases. This has not yet been tried, and would be an interesting experiment, although one lacking the same kinds of ambiguities that exist for gene names; a better subject might be synonyms of disease names from the UMLS Metathesaurus.

The technique relied on taking advantage of the fact that many Entrez Gene database entries have a field called the SUMMARY. The summary provides a concise statement of what is known about a variety of facets of the gene, such as its function, domains, splice variants, etc. (Jin *et al.* 2009). An example summary is

> TRPC1 belongs to the transient receptor potential (TRP) superfamily of cation channels. TRP cation channels are involved in diverse physiologic processes, including receptor- and store-operated Ca(2+) entry, mineral absorption, and cell death. They also function as sensors for pain, heat, cold, sound, stretch, and osmotic changes. (Zhang *et al.* 2009 [PubMed 19193631]) [supplied by OMIM] (http://www.ncbi.nlm.nih.gov/gene/7220)

Of the 9,390 genes in their document collection, 4,195 had a summary associated with them. This left them with two document collections: 9,390 with GeneRIFs, and 4,195 with GeneRIFs and SUMMARY fields. (A third document set is not discussed here.)

The technique has something in common with blind relevance feedback. The idea is that whenever a gene has a summary associated with it, the contents of that summary can be added to the name/symbol of the gene to expand the query. The beauty of the idea is that it allows for disambiguation of genes with ambiguous names.

The system was evaluated against two baselines. The first baseline consisted of the gene name and aliases ("official" name, "official" symbol, and known aliases) from the Entrez Gene record. The second baseline was those, plus the disjunction of the words *gene, genetics, genome*, and *oncogene*. These were added in an attempt to increase the ranking of documents from the genomics domain. The two baselines performed very similarly, and we will not differentiate between them further.

The authors experimented with two sources of additional knowledge for query expansion: fields representing the protein products of the gene, and the contents of the SUMMARY field. They also tried combining the protein products with the summary contents. In general, the protein product information added very little to performance, perhaps due to the rampant gene/product metonymy that we have discussed in Chapter 3. Adding information from the SUMMARY field turned out to be very successful in disambiguating gene names in information retrieval, with improvements of up to 17.7%, depending on the baseline.

5.4 A phrase-based search engine, with term and concept expansion and probabilistic relevance ranking

The Essie search engine (Ide, Loane, & Demner-Fushman 2007a) was originally developed in 2000 at the National Library of Medicine to support ClinicalTrials.gov, an online registry of clinical research studies. From the beginning, Essie was designed to use synonymy derived from the UMLS to facilitate consumers' access to information about clinical trials. Many consumers searching for medical information are unlikely to be familiar with the medical terminology used in the documents and use more common language in their queries. Most of the ClinicalTrials.gov documents about heart attacks do not contain the phrase "heart attack", but instead use the clinical term "myocardial infarction". Concept-based searching, which utilizes the UMLS-derived synonymy, has the potential to bridge this terminology gap.

Essie implements concept-based searching by expanding queries with synonymy derived from UMLS concepts. Essie includes phrase searches of the original text and inflectional variants in addition to concepts, thus it is less reliant on concept mapping, and should be more robust when concept mapping fails. Essie searches for phrases from the user's query by preserving word adjacency as specified in the query rather than indexing terms from a controlled vocabulary. Queries are further expanded to include a restricted set of inflectional variants, as opposed to many search engines that rely on stemming (Baeza-Yates & Ribeiro-Neto 1999).

Tokenization is another feature very important to successful retrieval in the biomedical domain, as demonstrated repeatedly in the TREC Genomics track evaluations. For example, the best average precision in the 2005 evaluation was achieved by a system that in addition to spaces defined the following places in which a string of text can be broken up: hyphen; between a letter and a digit; and between the lower case and upper case letters (Jiang & Zhai 2007). Much of Essie's success in the Genomics track 2003 evaluation can also be attributed to tokenization. Tokenization decisions in Essie are driven by characteristics of biomedical language in which punctuation is significant. Essie uses a fine-grained tokenization strategy. Every sequence of letters, sequence of digits, and individual punctuation characters are treated as separate tokens. For example, "non-hodgkin's lymphoma" consists of the six tokens:

1. non	2. -	3. hodgkin
4. '	5. s	6. lymphoma

The Essie search model consists of two distinct phases: indexing and searching. The indexing phase identifies and records the position of every token occurrence in the corpus. The tokenization process results in token adjacency indexes that contain position information about every token occurrence in the corpus. Two additional lookup tables are generated during the indexing process. The word variants dataset is derived from the corpus and the UMLS SPECIALIST Lexicon and is used for Term Expansion. The synonymy dataset is derived from the UMLS and is used for Concept Expansion.

The searching phase uses query expansions to produce a set of search phrases, each of which is a sequence of tokens. Tokens are matched against the indexes to identify potentially relevant documents. Then, probability of relevance is quantified by document scoring. Scoring is based on (1) the relative values of the different phrases produced by query expansion and (2) the relative values of the locations in the document where phrases were found. Essie's scoring algorithm can be summed up as preferring salient terms in valuable document fields. The salient terms are phrases from the query, with large penalties for dropping words or sub-phrases. There is less penalty for breaking word adjacency and small adjustments for using synonyms and word variants. The valuable fields of a structured document are determined by the nature of the document collection, for example, captions are the most important fields when searching for images.

5.5 Full text

One of the most exciting developments in biomedical information availability in the recent past has been the creation of PubMedCentral. While MEDLINE contains only abstracts, PubMedCentral makes the full texts of journal articles available.

This begs the question of whether or not full text is actually more effective than abstracts for finding relevant articles. Lin (2009) examined this question in detail. He found that examining the full text of journal articles per se did not yield an improvement in retrieval performance. However, when he tried using full text articles that had been segmented into paragraphs and returning articles just when relevant paragraphs were found within them, he did show an increase in retrieval effectiveness. Two facts bear explanation here – the failure of full articles to show an improvement in retrieval effectiveness, and the ability of subspans – i.e., paragraphs – from those articles to do so. Lin hypothesizes that the failure of full articles to show an improvement may be due to "noise" introduced by the additional content (that is, content in addition to what is found in the abstract and titles). For example, full articles might contain mentions of related work or future work that is not directly germaine to what a given article is about. Paragraphs, on the other hand, he speculates to repeat the important information in an article, perhaps in different ways. These different ways give more opportunities for terms in the query to match terms in the text. This idea is supported by the fact that the paragraph strategy specifically improves recall.

Table 5.1 Searching abstracts, full article bodies, and paragraphs from article bodies using the bm25 and Lucene algorithms. Values in parentheses are changes relative to the baseline. ** indicates statistically significant difference from the baseline at $P < .01$. * indicates statistically significant difference from the baseline at $P < .05$. Unmarked values are not statistically significantly different from the baseline. Data from Lin (2009).

MAP		
	bm25	Lucene
Abstract	0.163	0.129
Article	0.146 (−11%)	0.235 (+82%)**
Span	0.240 (+47%)**	0.206 (+60%)**
P20		
	bm25	Lucene
Abstract	0.322	0.293
Article	0.158 (−51%)	0.353 (+20%)*
Span	0.357 (+11%)	0.332 (+13%)
IPR50		
	bm25	Lucene
Abstract	0.110	0.090
Article	0.163 (+48%)	0.222 (+146%)**
Span	0.212 (+93%) **	0.159 (+77%)**

5.6 Image and figure search

One important difference between full-text articles and abstracts is that full-text articles contain images, figures, and tables, while abstracts do not. It has been observed that biomedical researchers often initially "read" journal articles by reading the abstract and then skimming through the article for the figures and captions (Sandusky & Tenopir 2008; Hearst *et al.* 2007a; Hearst *et al.* 2007b). There have been various attempts to make the process of retrieving images or of matching them to text easier. A first community-wide study of medical images retrieval was associated with the Cross Language Evaluation Forum 2004 Image Retrieval track. In this task, an example image was used to perform a search against a medical image database consisting of images such as scans and x-rays to find similar images. Each medical image or a group of images in the database represented an illness, and case notes in English or French were associated with each illness to be used for establishing diagnosis. Ad hoc retrieval of images as answers to text queries was introduced in ImageCLEFmed 2005, and in 2009 medical retrieval tasks were expanded to include case-based retrieval (retrieval of cases similar to a patient's case represented using text and images). The participating teams focus on retrieval methods rather than on development of user interfaces (few teams use the opportunity to submit interactive results). Several studies of user preferences for searching and viewing figures in journal articles were conducted by Hearst and colleagues (Hearst *et al.* 2007a; Hearst *et al.* 2007b; Divoli, Wooldridge, & Hearst 2010). The interface developed by Hearst *et al.* searches specifically within captions for all types of images, figures, and tables. A user study showed that a large majority of participants liked the idea and the interface and would be likely to use such a tool. We return to this study in the section on user interface evaluations (Section 10.6).

5.7 Captions

Research evidence indicates that augmenting MEDLINE citations with other relevant text can improve retrieval. For example, figure captions were instrumental in finding documents containing experimental evidence and discussing the Drosophila genes and their products (Regev *et al.* 2002). Regev *et al.* noticed that the evidence is often in the figures and used captions as substitutes. Shatkay, Chen, & Blostein (2006) examined the possibility of integrating information derived directly from image data with text for biomedical document categorization, and concluded that this method has potential. Also, Divoli, Wooldridge, & Hearst (2010) showed that captions should be searched.

5.7.1 Evaluation

Evaluations of the quality of document retrieval are one of the oldest and well-established areas of biomedical informatics research. It started with the 1969 evaluation of the 'on demand search' function of the Medical Literature Analysis and Retrieval System (MEDLARS) (Lancaster 1969). The study started with determining the prime requirements of demand search users that were assumed to be related to the coverage of the documents collection; recall, precision, and the response time of the system; the format in which search results were presented; and finally, the effort in query formulation needed to achieve a satisfactory response. Lancaster identified these two most critical problems in the evaluation: "1. Ensuring that the body of test requests was, as far as possible, representative of the complete spectrum of "kinds" of requests processed. 2. Establishing methods for determining recall and precision performance figures."

The principles of retrieval evaluations were later developed and tested in the Text REtrieval Conferences (Voorhees & Harman 2005). Due to the nature of test requests that reflected information needs of biologists and are based on interviews conducted with biologists working in industry, research and academia, the collections developed in the TREC Genomics track evaluations (2003–2007) are still a good source of reusable test collections. The TREC 2011 Medical records track attempted to create a reusable collection of de-identified patients' records and relevance judgments for cohort selection task. The information requests in this task consist of inclusion criteria for a cohort study, for example, finding patients who underwent robotic-assisted surgery. Due to difficulties in obtaining and distributing collections of patients' records, it is not clear if such evaluations will continue.

CHAPTER 6

Concept normalization

In practical biomedical text mining systems, it is often important to be able to map a mention of an entity in text to a specific entry in a database or other model of the world. One obstacle to this is lexical variability. For example, it might be important to recognize that *renal* and *kidney* tissue are the same and map statements about both to the same anatomical structure. Another obstactle is lexical ambiguity. For example, it is important to recognize that *PDA* can represent two distinct anatomical structures, the heart's posterior descending artery and an anomaly called a patent ductus arteriosus. The task has elements in common with word sense disambiguation and with general concept normalization.

Two especially important such areas in the biomedical domain are gene normalization and abbreviation definition.

6.1 Gene normalization

We have discussed the Entrez Gene database elsewhere in this book. Entrez Gene is an enormous database in which each entry is a specific gene from a specific species. It is often used as a standard reference in genomic research. One important task in biomedical text mining is extracting information about genes from text (see the chapter on relation extraction), and to be maximally useful, we must often know exactly which gene we are extracting information about. Homologous genes, or genes in different species that are descended from a common ancestor, may or may not have shared functions or other characteristics, and it is important to let the biomedical researcher decide for themselves when to consider them as the same or different.

Gene normalization can be defined narrowly as the task of taking a mention of a gene in text and mapping it to an identifier in a database. (A more specific definition of gene normalization coming from the BioCreative shared tasks will be given below.) Multiple factors conspire to make it difficult. First, of course, is that to be able to normalize a gene mention, we must first recognize it as a gene mention. The chapter on named entity recognition discussed in detail what makes this difficult. Next, there may be discrepancies between the way that the gene name is given in the database and the way that it appears in the text – *BRCA1* may appear as *BRCA1, brca1, BRCA-1, brca 1,* or other variants. Even if we have managed to recognize the gene mention and normalize it to a form that actually appears in the database, that form may appear in the database in multiple entries. For instance, BRCA1 is the name of a gene in many species, and TRP1 is the name of five different genes just in the human species alone. So, practical gene normalization comprises a number of problems – gene mention, term normalization, and a sort of word sense disambiguation.

The gravity of the problem can be seen from looking at the results of shared tasks in which people have tried to do information extraction tasks that included gene normalizations. The first of these was in the first BioCreative evaluation. This evaluation included a task in which participants were to map Gene Ontology terms (see Chapter 7) to specific SWISS-PROT entries. (SWISS-PROT is a manually curated database of information about proteins in many different species.) Performance of all participants was quite low, and while it was clear that the contributions to this low performance included the difficulty of recognizing Gene Ontology terms and the difficulties inherent in any relation extraction task, it was also clear that a major obstacle was mapping the extracted Gene Ontology terms to the correct protein. We see the gravity of the gene normalization task again in the BioCreative II and BioCreative II.5 protein–protein interaction tasks. Both evaluations required participants to find assertions about interacting proteins and map them to specific UniProt identifiers. (UniProt is a superset of SWISS-PROT which includes data that is not manually curated.) While the problem here is once again conflated with the challenges of the relation extraction task, protein name normalization was again a clear contributor to low performance numbers in the 30s.

The difficulty of the gene normalization task has been seen to vary considerably between species. How this variation plays out is, however, best understood in a case with some simplifying assumptions (such as the assumption that the species under discussion in the document is already known); the effect of species on gene normalization difficulty in more realistic task settings has not yet been studied.

6.1.1 The BioCreative definition of the gene normalization task

Although issues in gene normalization were studied as early as Cohen *et al.* (2002), the task was given an influential definition by the BioCreative shared tasks. Although this definition was somewhat simplified and, in an important respect, artificial, it was highly influential and led to a number of insights into the nature of gene normalization and should be understood.

On the BioCreative definition of the gene normalization task, one is given:

- A database of genes for a specific species
- A set of documents that contain mentions of genes from that species

... and the task is to return, for each document, the list of identifiers of all genes from that species that are mentioned in the document.

To accomplish this task, every gene that is mentioned must have a mention detected at least once. At the same time, it must be recognized when surface-different mentions actually refer to the same gene. For example, the system must not return separate identifiers for *Est6* and *esterase 6*, but must recognize that they are two different forms of the same gene name.

Unlike the gene mention task, there is no requirement to return a specific text span in the BioCreative task definition. We return to this below.

The BioCreative definition of the task is a simplified abstraction in two ways. The most obvious is that the system is provided a priori with the knowledge of what the species is. This means that any genes with the same name in other species can simply be ignored, and the only disambiguation problem is with other genes with the same name in the same species.

The other simplification is that genes are identified on the document level, rather than on the level of specific occurrences in text. This is relevant because information extraction requires identification on the level of the individual mention – identification at the level of the document is irrelevant for this application.

However, despite the specific mention issue, the BioCreative definition of the task is still useful for information retrieval applications, where document-level identification is sufficient.

6.2 Building a successful gene normalization system

Any gene normalization system will at some point consult a dictionary, and at this point will have to deal with the fact that any gene name may be written in a number of different ways. Thus, it is likely the case that all gene normalization systems include a step in which the typographic form of names in the dictionary and in the text are normalized (in the simplest sense of that term) to the same form. Cohen *et al.* (2002) investigated the effects of a number of such simple normalization steps on matches to the Entrez Gene database (then known as LocusLink), and most gene normalization systems take advantage of those techniques, implicitly or explicitly, as well as some others. Common variations that need to be normalized away include:

- Case
- Presence of hyphens or spaces
- Prefixes indicating species
- Abbreviated or spelt-out forms of Greek letters
- Equivalence of Greek and Roman letters
- Equivalence of letters and numbers (e.g. *A* with *1*)
- Occasional equivalence of *A* or *1* with no number or letter at all

Successful gene normalization systems have in common the use of knowledge. On the simplest approaches, this might be purely lexical knowledge applied in a Leskian fashion (Lesk 1986) to disambiguate between ambiguous names. Where systems like this typically differ is in where they look for the reflection of lexical knowledge. For example, they might look very locally, e.g. in adjacent appositive parenthesized definitions, or within the same sentence. They may look further afield in the document, e.g. throughout the abstract. What is used as the proxy for a Leskian definition, i.e. the source of lexical knowledge, can also differ. For example, a system might restrict itself to the gene name field in Entrez Gene, or may look at some larger but definition-sized chunk of text like the Entrez Gene SUMMARY field, or may look much further afield, at documents that are linked to the candidate gene entries, possibly with tf*idf weightings (Neves, Carazo, & Pascual-Montano 2010).

More complicated approaches make use of a wider array of types of knowledge. For example, the GNAT system (Hakenberg *et al.* 2008, 2011) makes use of mentions in the text of chromosomal locations, protein lengths, cell types, domains, interaction partners, and Gene Ontology terms that are mentioned in the article and are also available in databased form.

6.2.1 Coordination and ranges

One important contributor to the performance of high-performing systems has been the ability to handle coordination, whether it be simple coordination, ranges of numbers or letters at the end of names, or other related phenomena. Baumgartner *et al.* (2008) found that 8% of gene mentions in the BioCreative II data set involved coordinations. Baumgartner Jr. *et al.* (2008) and Lu (2007) identified a number of forms of coordination and ranges of gene names that appear in biomedical text.

- Coordination of numbers or letters at the end of names, e.g. *HMG1 and 2* (PMID 9600082)
- *IL3/5*, meaning *IL3* and *IL5*
- *freac1-freac7*, meaning *freac1*, *freac2*, etc., through *freac7*

- *IL3-5*, meaning *IL3* and IL5
- *M and B creatine kinase*, meaning *M creatine kinase* and *B creatine kinase* (PMID 2364108)

6.2.2 An example system

As an example of a successful system that makes clever use of external knowledge sources, consider the GNAT system (Hakenberg *et al.* 2008). GNAT attempts to do species-independent gene normalization – that is, it attempts to normalize genes without being given a priori knowledge about what species they are in (see Figure 6.1).

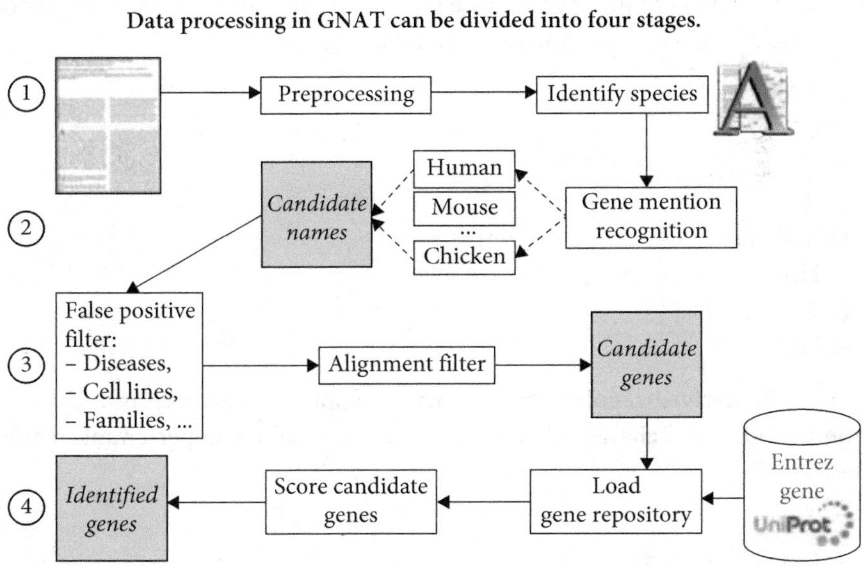

Figure 6.1 Data processing in GNAT can be divided into four stages. (1) After preprocessing a text, we search for species using AliBaba. (2) For each species that is referenced in the text, a dictionary is loaded and we annotate names of genes in the text. (3) Filters remove false positives, for example, names of gene families and diseases. (4) Remaining candidate genes are ranked using context profiles stored in a repository created from EntrezGene and UniProt annotations. (Hakenberg *et al.* 2008)

Prior to run-time, a set of dictionaries is assembled on a per-species basis. Each dictionary is then filtered to remove non-gene names from gene name fields. (This is actually a pervasive problem – gene name fields may contain entries like *fragment* (GeneID 4177514), *hypothetical protein* (GeneID 5830418), *hypothetical*

(GeneID 100144602), and the like.) They also removed names which seemed unlikely to have matches in text due to not fitting normal English language patterns or containing large amounts of parenthesized material[8]. Using an unspecified technique, they also removed names which did not seem to fit in with the other synonyms for a gene – for example, in the classic *esterase 6* example, they would have removed *Est5*.

The remaining names are then converted to regular expressions, using a complicated procedure that separates gene names into chunks, turns the individual chunks into regular expressions, and then unifies the smaller regular expressions into one single regular expression as a whole. A new chunk is made whenever there is a change in case (except when the switch is from upper to lower case, and it is at the beginning of a word, and the chunk length is greater than or equal to three), a switch from letters to numbers or vice versa, or a symbol. Thus, for the symbols *CD95R* and *Hfn-3beta*, you would get the following chunkings:

- CD95R
  - CD
  - 95
  - R
- Hfn-3beta
  - Hfn
  - 3
  - beta

Heuristically derived regular expressions are applied to each chunk, depending on its individual characteristics. They yield the following per-chunk regular expressions:

- CD95R
  - CD: (CD|Cd|cd)
  - 95: 95
  - R: (R|r)(eceptor)?
- Hfn-3beta
  - Hfn: (HFN|Hfn|hfn|HfN)
  - 3: (3|III|iii)
  - beta: (B|b)(eta)?

8. The data in Cohen *et al.* (2002) suggests that parenthesized material in gene name dictionaries can safely be discarded.

As can be seen from the different handling of *95* in *CD95R* and *3* in *Hfn-3beta*, the heuristics are indeed finely tuned to the individual characteristics of the chunks.

These per-chunk regular expressions are then unified so as to allow for whitespace, hyphens, or no whitespace or hyphen between each chunk, yielding

- CD95R: (CD|Cd|cd)[-s]?95[-s]?(R|r)(eceptor)?
- Hfn-3beta: (HFN|Hfn|hfn|Hfn)[-s]?(3|III|iii)[-s]?(B|b)(eta)?

The system then examines the input text to determine which species are mentioned in it. Two tactics are used – looking for species names, and looking for cell line identifiers. Cell lines are populations of cells that are bred from a common ancestor population. Since for every cell line we know which species the original ancestor population comes from, cell line identifiers are good indicators of species.

The next step taken is to find gene mentions. Once these have been found, individual gene mentions are mapped to a species. Recall that all species mentioned in the article were identified in an earlier step. Those species are then mapped to the individual gene mentions using the following rules, in order:

- species is mentioned within the noun phrase
- species is mentioned in the same sentence
- species is mentioned in the previous sentence
- species is mentioned in title of article
- species is mentioned in the first sentence of the abstract
- species is mentioned anywhere in the abstract
- species is annotated as a MeSH term for the article

This yields a mapping between a mention and a species, but may still not map to the appropriate gene. If only one gene in the species has a name that is a potential match, then the job is done. However, there may be multiple genes in the species that map to that name – consider the case of TRP1 in humans, which as noted earlier is a symbol or synonym for five different genes. Now the system makes use of other information available for genes. In particular, the system uses:

- EntrezGene summaries
- GeneRIFs
- chromosomal location
- diseases
- functions
- tissues in which the gene is known to be expressed
- keywords
- protein length

- protein mass
- known mutations
- domains
- interaction partners
- Gene Ontology terms

A score is assigned for each match to a piece of this background information. Every gene is then assigned a score that is the sum of these individual scores. The gene with the highest score is then returned by the system.

As an example, consider the following input:

> The P54 gene was previously isolated from the chromosome translocation breakpoint region on 11q23 of RC-K8 cells, with t(11;14)(q23;q32). It was found to encode a 472-483-amino-acid (aa) polypeptide belonging to an RNA helicase/translation initiation factor family. (PMID 8543178)

RC-K8 is the identifier of a human cell line, so we can immediately restrict our search to human genes with the name P54. P54 is an alias for five human genes, so we must go to the step of using the background knowledge about these genes to select the correct one. The information about those genes is as given in Table 6.1. Of those genes, three score zero, one scores one, and one scores three. That is DDX6, which is returned as the correct mapping.

On a set of 100 abstracts containing 34 species, the system scored an F-measure of 81.4. This is comparable to the performance of many systems on single-species data sets.

Table 6.1 Background information for the three human candidates for P54 (Hakenberg *et al.* 2008).

Gene	**DDX6**	ETS1	FKBP5	NONO	SRFS11
Chromosome	e **11q23.3**	textbf11q23.3	6p21.3–p21.2	X113.1	1p31
length	**483**	441	457	471	484
Gene Ontology terms	**RNA helicase activity**	immune response	FK506 binding	DNA binding	mRNA processing

6.3 Normalization and extraction of clinically pertinent terms

Finding and normalizing clinical terms is essential in such tasks as linking research evidence found in the literature to patients' Electronic Health Records, biosurveillance, quality assurance and many others. "To facilitate the development of conceptual connections between users and relevant machine-readable information" (Humphreys & Lindberg 1993) the National Library of Medicine developed the Unified Medical Language System (UMLS). The system's knowledge sources, the SPECIALIST Lexicon and the Metathesaurus, provide a means to establish conceptual connections between users and information sources. We will describe MetaMap, a system that uses these resources to provide the most comprehensive (at the time of this writing) mapping and normalization of free text (both clinical narrative and scientific publications) (Aronson & Lang 2010).

6.3.1 MetaMap UMLS mapping tools

MetaMap, designed to find Metathesaurus concepts in biomedical text, was developed and is maintained at the NLM (Aronson 2001b). Although several other tools have been developed for the purpose of mapping text to concepts in the UMLS Metathesaurus in the past (Zieman & Bleich 1997; Denny *et al.* 2002; Zou *et al.* 2003), only MetaMap attempts identifying all UMLS concepts in any document type, as well as determining if the identified terms are negated. Many MetaMap parameters are configurable. For example, a user can select one of the three data models that differ from each other by the level of filtering they do on the UMLS Knowledge Sources. The strict model is considered to be most appropriate for most applications. MetaMap first uses a minimal commitment parser to break the text into phrases. For each phrase, MetaMap generates acronyms, abbreviations, synonyms, derivational, inflection and spelling variants using the SPECIALIST lexicon and a list of synonyms. (The default set of variants is restricted to spelling variants, inflectional morphemes, derivational variation between nouns and adjectives, and synonyms.) The variants are used to retrieve a set of Metathesaurus strings called candidates. Each candidate is evaluated for the strength of mapping to the original text. The strength score is an average of four metrics: centrality, variation, coverage and cohesiveness. Centrality is a binary value, which equals 1 if a Metathesaurus string matches the head of the noun phrase, and 0 otherwise. Variation is computed as inverse edit distance between a Meta string and the noun phrase, where edit distance for spelling variants is 0, for inflectional variants is 1, for synonym or acronym/abbreviation variants is 2, and for derivational variants is 3. Coverage value indicates how much of the Metathesaurus string and the phrase are involved in the match. The cohesiveness value is similar to the coverage value but, unlike coverage, it does not ignore gaps in phrases and strings. Candidates with the best scores are selected as final mappings.

MetaMap performance in terms of its ability to find concepts was evaluated in a small scale study using 133 unique reference concepts identified by six people in 60 titles of medical articles (Pratt & Yetisgen-Yildiz 2003). Of the 133 concepts only 73 were in the UMLS Metathesaurus. Under lenient conditions, where a MetaMap identified concept was considered a match if at least one subject also identified the concept, MetaMap achieved 93.3% recall and 84.5% precision.

There are several sources of MetaMap errors. UMLS coverage and ambiguity are external to MetaMap processing, but influence its results significantly. Ambiguity in processing of biomedical publications can be somewhat alleviated by the built-in tool that disambiguates mappings using the context (Humphrey *et al.* 2006). Accepting or rejecting a candidate depending on its context might be helpful even if only one mapping is available, as in the following example:

```
Processing phrase: "with a concise description"
Meta Mapping (888):
    694 Concise [Biomedical or Dental Material, Organic Chemical]
    861 description [Research Activity]
```

The dental material Concise is the only candidate for the phrase, however of 4894 PubMed abstracts containing the word *concise* only 131 are about *bisphenol a-glycidyl methacrylate*, the chemical named Concise.

Another source of errors is overmatching, for example, the phrase: "aided" was mapped with a high score to:

```
AIDS (Acquired Immunodeficiency Syndrome) [Disease or Syndrome]
Aid <2> (Manufactured aid) [Manufactured Object]
```

In some cases a correct concept is generated as a candidate phrase, but is ranked lower than other candidates, and therefore not available in final mappings. For example, the phrase: "prognostic value" had a candidate score 623 as

```
Prognosis <1> (Forecast of outcome) [Health Care Activity]
```

but 694 as prognostic [Intellectual Product], which was retained as a final mapping.

The MetaMap output shown above is "human" readable, and loses some information. The XML or the original MetaMap machine output retain all information generated by MetaMap.

Recent research has indicated that the MGREP tool may be more effective than MetaMap for at least some categories of semantic concepts (Bathia *et al.* 2008), but overall MetaMap remains the industry standard for recognizing medical concepts in text, and it has been shown to be applicable to the biological domain as well (Aronson *et al.* 2005b) and to a wide range of task types, ranging from information retrieval to information extraction and literature-based discovery.

The development of the tool in Prolog started in 1991. MetaMap processing of the input text starts with lexical syntactic analysis that includes:

- tokenization, sentence boundary determination and acronym/abbreviation identification;
- part-of-speech tagging;
- lexical lookup of input words in the SPECIALIST lexicon;
- a final syntactic analysis consisting of a shallow parse in which phrases and their lexical heads are identified by the SPECIALIST minimal commitment parser.

Each phrase found by this analysis is further analyzed to generate variants by table lookup, identify matching Metathesaurus strings, and find the longest best-matching one. The set of concepts represented by the best matching strings can be further reduced using MetaMap's word sense disambiguation (Humphrey *et al.* 2006). The best matching concepts: (1) include the linguistic head of the input text; (2) have little or no variation between all input text words and their matching candidate words; (3) involve all or a significant portion of the input text in the mapping; and (4) include many contiguous chunks of the input text.

MetaMap employs two data models, relaxed and strict, that differ in how much Metathesaurus content is filtered out. The relaxed model filters out lexically similar strings based on case and hyphen variation, possessives, comma uninversion, NOS (Not Otherwise Specified) variation[9] and non-essential parentheticals. It also includes the manual removal of some strings such as numbers, single alphabetics, NEC (Not Elsewhere Covered) terms, Enzyme Commission (EC) terms, the short forms of brand names and, most importantly, unnecessarily ambiguous terms, such as *prostate* being synonymous with *prostate cancer*. MetaMap's strict model also filters out strings with complex syntactic structure; these are strings which MetaMap does not match well anyway. Table 6.2 presents examples of removed strings. Over 40% of Metathesaurus strings are removed in the creation of the strict model. It is MetaMap's default model for semantic NLP processing, and it has been available as the first Metathesaurus Content View since the 2005AA UMLS release (Demner-Fushman *et al.* 2010).

9. For practical purposes, the information content in the terms containing the NOS qualifier is hard to distinguish from that in the representations without the qualifier, for example, the meaning of the term Other myeloid leukemia NOS (CUI: C0029670) is not significantly different from that of Myeloid Leukemia (CUI: C0023470).

Table 6.2 String filtering in the MetaMap strict model.

Metathesaurus strings	Reason(s) for removal
Intraductal carcinoma, non-infiltrating NOS (morphologic abnormality)	NOS variation, comma uninversion, parenthetical, case, hyphen
[D] Castleman's disease (disorder)	Parenthetical, case, possessive
[M]Hodgkin's sarcoma	Parenthetical, case, possessive
[X]Diffuse non-Hodgkin's lymphoma, unspecified (disorder)	Comma uninversion, parenthetical, case, hyphen, possessive

MetaMap is configurable across multiple dimensions, including:

- data options, which choose the vocabularies and data model to use;
- output options, which determine the nature and format of the output generated by MetaMap;
- processing options, which control the algorithmic computations to be performed by MetaMap.

The data options determine the release of the UMLS Metathesaurus and whether the strict or relaxed model will be used. The output options allow the user to request machine or human-readable format; excluding or restricting to concepts of specified semantic types; and excluding or restricting to specified vocabularies. The processing options include: controlling the types of derivational variants used in lexical variant generation (no variants at all, adjective/noun variants only, or all variants); using the WSD module; processing each record as a single phrase; allowing overmatches (retrieving all concepts containing the input phrase, for example, using the overmatch option for a text containing the phrase "skin" generates almost 40, 000 candidate mappings contained in the 2009 strict data model). The candidates range from "Skin (Entire skin) [Body System]" to "Derma-Gard 32-291-06 skin wipe [Pharmacologic Substance]".) Another processing option allows concept gaps so that, for example, the text "obstructive apnea" will map to concepts "obstructive sleep apnoea" and "obstructive neonatal apnea". Details of all aspects of MetaMap processing can be found in the technical documents at the MetaMap portal (http://mmtx.nlm.nih.gov/).

MetaMap is ideally suited for off-line processing, as evidenced by its use in generation of MEDLINE indexing recommendations. Its complexity and thoroughness make it less suitable for online applications. MetaMap is highly configurable (even its lexicon and target vocabulary can be replaced with others from another domain). MetaMap can be easily customized for a given task; however, customization requires thoughtful choice of options. Prior to releasing MetaMap as open source software and providing Java API and UIMA wrappers for MetaMap output, NLM maintained an open source Java-based version of MetaMap, called MMTx (MetaMap Transfer.) MMTx is currently being phased out in favor of the Prolog version that is now much faster. In addition, there was always about a 20% difference betwen MetaMap and MMTx in the overall results, and maintaining versions of the same tool in two different languages proved to be impractical. Nevertheless, the datasets for MMTx are still updated and the software is available for download from http://mmtx.nlm.nih.gov/MMTx/.

CHAPTER 7

Ontologies and computational lexical semantics

One of the crucial enabling factors in the process of genome-level science over the past few years has been the development of ontologies of the biomedical domain. An ontology can be defined as a set of:

- Concepts, typically with terms and unique identifiers associated with them, and sometimes with synonyms
- Relations between the concepts, typically is-a and has-part
- Definitions of the concepts

By design, it is the definition, not the term, that constitutes the concept. So, if a term changes, there is no change to the concept as a whole. However, if a definition changes, then the concept identifier is deprecated and a new concept is created.

Meanwhile, in medicine, ontology-like groups of concepts have been used for years for purposes ranging from billing to indexing the biomedical literature.

7.1 Unified Medical Language System (UMLS)

The development of knowledge-intensive methods and tools in the medical domain is made possible by the Unified Medical Language System[10]. The UMLS, maintained at the National Library of Medicine, consists of three knowledge sources: Metathesaurus, Semantic Network, and SPECIALIST Lexicon. The UMLS Metathesaurus contains information about biomedical concepts, their various names, and the relationships among them. It represents many source vocabularies (thesauri, classifications, code sets, and lists of controlled terms) in a single database format with the purpose of linking alternative names and views of the same concept together. The Metathesaurus preserves the names, meanings, hierarchical contexts, attributes, and inter-term relationships present in its source vocabularies. It also establishes new relationships between terms from different source vocabularies. The concept *Lou Gehrig's Disease* illustrates linking of a concept name to its synonyms. Through its Metathesaurus unique concept identifier (CUI = C0002736) this string is recognized as synonymous with the following terms:

10. http://www.nlm.nih.gov/research/umls/

- Amyotrophic Lateral Sclerosis
- ALS
- ALS (Amyotrophic Lateral Sclerosis)
- ALS - Amyotroph lat sclerosis
- Amyotrophic Lateral Sclerosis/Progressive Muscular Atrophy
- Amyotrophic lateral sclerosis (disorder)
- amyotrophy; lateral sclerosis
- Bulbar motor neuron disease
- Gehrig's Disease
- Lou Gehrig Disease
- Motor Neuron Disease, Amyotrophic Lateral Sclerosis
- palsy; creeping
- spinal; sclerosis, lateral (amyotrophic)

The 14 synonyms were found in 19 source vocabularies (Alcohol and Other Drug Thesaurus, Clinical Problem Statements, COSTAR, CRISP Thesaurus, DXplain, ICD-9-CM, ICPC2-ICD10 Thesaurus, Library of Congress Subject Headings, MedDRA, MedlinePlus, MeSH, UMLS ICD-9-CM Terms, NCI Thesaurus, National Drug File – Reference Terminology, Quick Medical Reference, Read Codes, SNOMED 1982, SNOMED Intl 1998, and SNOMED Clinical Terms).

The 2010AA version of the UMLS Metathesaurus contains information about over 2.2 million biomedical concepts and 8.2 million concept names from more than 150 source vocabularies. The exact metadata for each Metathesaurus release are provided in the metadata files, for example:

MRFILES	The names and sizes of every Metathesaurus file
MRCOLS	The names and size range of every Metathesaurus data element
MRSAB	The source vocabularies in the Metathesaurus

Realizing that the Metathesaurus must be customized to be used effectively, NLM provides customization tools and online vocabulary services.

The UMLS Semantic Network categorizes each Metathesaurus concept into at least one of the basic semantic types. It also defines the set of relationships that may hold between the semantic types. The 2010 release of the Semantic Network contains 133 semantic types and 54 relationships. There are major groupings of semantic types for organisms, anatomical structures, biological function, chemicals, events, physical objects, and concepts or ideas. The Semantic Network provides textual descriptions of semantic types and defines important relations in the biomedical domain, in addition to 54 relationships allowed between the semantic types. The primary relation between the semantic types is "IS-A". It establishes the hierarchy of types within the Network and is used for deciding on the most specific semantic type available for assignment to a Metathesaurus concept. There is also a set of non-hierarchical relationships, which are grouped into five major categories:

a. physically related to
b. spatially related to
c. temporally related to
d. functionally related to
e. conceptually related to

Applied to the *Lou Gehrig's Disease* example, the Semantic Network categorizes it as "Disease or Syndrome," and provides several hierarchies (see Figure 7.1) that contain this disease and related concepts, e.g. the ALS3 gene.

The SPECIALIST lexicon is a general English lexicon that includes biomedical terms. The lexicon entry for each word or term contains the syntactic, morphological, and orthographic information. Lexical entries may be single or multi-word terms. Each lexical record has a base form, a part of speech, a unique identifier and optionally a set of spelling variants. The base form is the uninflected form of the lexical item; the singular form in the case of a noun, the infinitive form in the case of a verb, and the positive form in the case of an adjective or adverb. Lexical information includes syntactic category, inflectional variation (e.g., singular and plural for nouns, the conjugations of verbs, the positive, comparative, and superlative for adjectives and adverbs), and allowable complementation patterns.

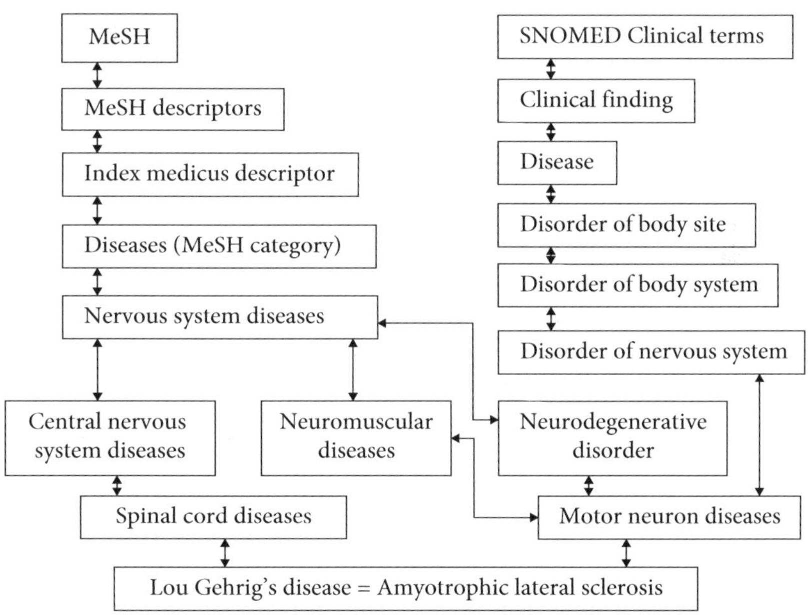

Figure 7.1 The concept "Lou Gehrig's Disease" and some of its hypernyms.

7.1.1 The Gene Ontology

In the genomics world, one of the most influential ontologies has been the Gene Ontology (The Gene Ontology Consortium 2000; Consortium 2001). The Gene Ontology was invented when researchers noted that they had problems answering a very basic type of question: does this gene in Organism A have the same function as this gene in Organism B? Terminological differences in genomic databases of different species made it difficult to answer these questions on any sort of large scale. Thus, scientists from three model organism databases joined together to create an ontology of gene functions. From the perspective of the ethnography of science, one of the interesting things about this resource is that it was undertaken without recourse to ontologists or linguists. (Both ontologists and linguists have since become involved in the Gene Ontology.)

Today the Gene Ontology (GO) project, representative of specialized knowledge sources, is a collaboration of seventeen organizations developing consistent species-independent descriptions of gene products in different model organism databases. The Gene Ontology contains three structured controlled vocabularies that describe biological processes associated with gene products; cellular components; and molecular functions. Cellular components are defined as parts of some larger object including anatomical structure (for example, cell nucleus) or a gene product group (for example, proteasome). Biological processes are defined as series of events, for example, signal transduction. The distinction between a biological process and a molecular function is that a process must have more than one distinct step. Biological process is not equivalent to a pathway either, because the description of a process does not include dynamics or dependencies required to fully describe a pathway. Molecular function terms describe actions performed by individual gene products at the molecular level, for example, catalytic activity or binding.

7.2 Recognizing ontology terms in text

Recognizing concepts from ontologies when they occur in free text has been identified as a potential major contribution of natural language processing to a variety of fields, ranging from medical billing to construction of genomic databases. As noted above in the definition of *ontology*, a concept in an ontology typically has a term associated with it. One direct approach to recognizing mentions of concepts is through recognition of these terms. (Indirect approaches require inference or external knowledge.)

Ontology terms can be extremely difficult to recognize in text. This is because of both linguistic variability and the necessity of domain knowledge. For example, the following seven variants of the term *leukocyte cell migration* (GO identifier GO:0050900) have all been observed in actual text:

- *leukocyte migration* (PMID 8992620)
- *migration of leukocytes* (PMID 10577496)
- *leukocytes migrate* (PMID 8200044)
- *migrating leukocytes* (PMID 18755005)
- *eosinophil migration* (PMID 10534125)
- *neutrophils migrated* (PMID 2961775)

The first four examples all require recognizing syntactic or morphological variability; the last two examples require the knowledge that eosinophils and neutrophils are types of leukocytes, as well as linguistic variation.

Even recognizing exact matches to terms may present problems. For example, tokenization at an earlier processing step may cause a simple dictionary look-up to fail. Subtler problems may exist, as well. For example, Cohen *et al.* (2010c) examined the use of a structured test suite in evaluating the performance of a system for recognizing exact matches of ontology terms and found that it failed on Gene Ontology terms that contained the word *in*, although not on other stopwords or prepositions.

Overall, very few exactly matching Gene Ontology terms actually occur in text. Verspoor, Joslyn, & Papcun (2003) examined a corpus of 2.3 million words and found that only 6% of full Gene Ontology terms occurred in their official forms, of which a little more than half were multi-word phrases.

7.3 NLP for ontology quality assurance

Quality assurance of ontologies is, in general, an understudied field with little consensus about how it should be done and a great need for practical techniques. Verspoor *et al.* (2009) demonstrated that it is possible to use natural language processing techniques for ontology quality assurance. They tackled the specific problem of violations of univocality. In this context, *univocality* refers to the consistent expression of relations or subconcepts in the terms that are attached to concepts. Gene Ontology curators identified violations of univocality as a major problem in ontology quality assurance, for example potentially leading to redundant terms such as *regulation of transcription* and *transcription regulation*.

The basic approach taken by Verspoor *et al.* (2009) was to apply a small set of transformations to Gene Ontology terms, and then cluster together terms that transformed to a similar shape. (A similar technique was applied by Cohen *et al.* 2002 to gene names from LocusLink [a precursor to Entrez Gene] and resulted in the finding of a number of errors in gene names.) The three transformations applied were:

- Replacement of embedded Gene Ontology and ChEBI terms with a constant string. For example, the GO term *regulation of cell motility*, where *cell motility* is itself a GO term, becomes *regulat of GTERM*.
- Removal of stopwords.
- Ordering words in the term alphabetically.

Unlike the transformations in Cohen *et al.* (2002), which were primarily string-based, these transformations were meant to identify semantically similar terms. Stopword removal consists of removing semantically contentless words. Term replacement aims at semantic abstraction.

After applying the transformations, similarly-shaped transformed terms were clustered together, and then a heuristic search was applied to identify potential univocality violations.

Ogren, Cohen, & Hunter (2005) showed that concepts from descriptive linguistics could be used to find missing relations between terms in ontologies. Using a discovery procedure from descriptive linguistics, they looked for Gene Ontology terms that contained other Gene Ontology terms as substrings. They then modelled the added strings (i.e. the material added to the substring) as derivational morphemes by looking for derivational strings (strings that derived new Gene Ontology terms from other Gene Ontology terms) that always occurred under the same subontology of the Gene Ontology. In this way, they were able to discover a number of terms that should have been related to other terms but were not, such as *proliferation* and *regulation of cell proliferation*. These findings were utilized to enrich the set of linkages, and later the set of relations, in the Gene Ontology.

7.4 Mapping, alignment, and linking of ontologies

As the number of ontologies has grown, there has been increasing interest in relating them to each other. A common use case is to incorporate an ontology into the UMLS. Another is to find equivalent concepts in different ontologies. This is of use when one's goal is to keep ontologies completely orthogonal, as is the case with the OBO ontologies (a set of ontologies that commit to basic structural and organizational principles endorsed by the National Center for Biomedical Ontology). Another motivation is to increase support for automated reasoning.

The terms *mapping, alignment,* and *linking* of ontologies are not well-defined in the literature. However, at a first approximation, we can define them as follows:

- *Mapping* is the detection of identical concepts in different ontologies.
- *Alignment* is the detection of meaningful relationships between different concepts in different ontologies or within the same ontology.
- *Linking* is the detection of any relationship between different concepts in the same or different ontologies.

Two broad families of methods have been proposed for the mapping, alignment, and linking of ontologies (MALO). One family of approaches utilizes the structure of the ontologies. The other family of approaches, on which we will focus here, utilizes what might be termed lexical methods. *Lexical methods for MALO* consist of methods that make use of linguistic information. This information is typically found in the terms that constitute concept names, although definitions have also been used to detect relationships between concepts.

Johnson *et al.* (2006) evaluated a number of lexical methods for MALO. They also looked at the relative efficacy of terms and of definitions as input to the MALO algorithm. Their goal was to find relationships between the BRENDA tissue ontology, the Cell Type Ontology, and the ChEBI chemical ontology, on the one hand, and the Gene Ontology on the other. Their basic methodology was to look for mentions of concepts from the first three ontologies within concepts in the Gene Ontology. As their methodology for finding such mentions, they used the Lucene search engine, modelling the concepts of the Gene Ontology as a collection of documents and modelling terms from the other three ontologies as queries. They evaluated three specific linguistic approaches to detecting hits:

- Exact match
- Using synonyms
- Stemming and stop word removal

In addition, they contrasted the effectiveness of searching only within terms versus searching within definitions as well.

Altogether, 91,385 apparent relationships were found between the three ontologies and the Gene Ontology. Of these, a random sample of 2,389 putatively associated pairs of terms were manually examined by two domain experts (inter-annotator agreement 93.5% before dispute resolution and 98.2% after dispute resolution).

The overall results can be found in Table 7.1. The first thing to note is that no approach was perfect – even exact matches yielded some errors, due to polysemy.

For example, the word *group* is a ChEBI term (CHEBI:24433), but is also a General English word and often appears with its General English sense in Gene Ontology definitions. *Joint* is a BRENDA term, with the meaning of anatomical joint, but often appears with the General English sense *combined* in Gene Ontology terms. Even domain-specific, non-General-English terms exhibited this behavior.

Table 7.1 Correctness rates (correct/evaluated) for lexical MALO techniques (Johnson *et al.* 2006).

Ontology	Exact	Synonyms	Stemming
ChEBI			
GO Term	99.5% (199/200)	42.0% (42/100)	73.0% (73/100)
GO Def.	97.8% (451/461)	69.0% (138/200)	74.0% (74/100)
Cell Type			
GO Term	100% (200/200)	94% (44/47)	76% (41/54)
GO Def.	98.7% (231/234)	50% (21/42)	92% (47/51)
BRENDA			
GO Term	76.0% (76/100)	83.0% (83/100)	7.0% (7/100)
GO Def.	93.0% (93/100)	69.0% (69/100)	15.0% (15/100)

A second thing to note is that different lexical techniques exhibited different performance characteristics for different pairs of ontologies. For example, exact matching of Cell Type terms against Gene Ontology terms was 100% accurate (200 correct out of 200 evaluated), but exact matching of BRENDA tissue ontology terms against Gene Ontology terms was only 76% accurate (76 correct out of 100 evaluated).

Adding synonyms to the query string resulted in an overall increase of 36% in the detection of apparent relationships between concepts. For example, it enabled matching the BRENDA term *adipose* (BTO:0000441) to the Gene Ontology term *larval fat body development* (GO:0007504). However, we have seen in information retrieval in general that the increase in matches from using synonym expansion comes at a price due to the concomitant increase in opportunities for polysemy, and not surprisingly, this technique had widely varying results, ranging from a high of 94% for matching the Cell Type Ontology against Gene Ontology terms to a disconcerting 42% for matching ChEBI terms against Gene Ontology terms. Overall, the accuracy of query expansion using synonyms was only 67.4%.

The final approach that was evaluated conflated stemming and stopword removal. This was done using an out-of-the-box Lucene analyzer called PorterStemmer. This technique found the largest number of relationships – for example, matching the Cell Type Ontology term *neuron* (CL:0000540) to the word *neuronal* in the definition of the Gene Ontology concept *syntrophin* (GO:0016013) – but had the lowest accuracy – the average correctness rate was only 51%. (This number is actually artificially low, as will be explained momentarily.)

A novel aspect of this work was that it evaluated the efficacy of using concept definitions, rather than just terms, for MALO. The authors found that searching definitions yielded a large increase in the number of relationships detected, with accuracies comparable to those for searching just in terms.

A number of conclusions about lexical methods for MALO can be drawn from this work:

- No method performs 100% correctly. Although exact matching was generally correct, in the case of mapping BRENDA concepts to GO terms it achieved an accuracy of only 76% percent.
- Most of the relationships detected by these methods are indirect. For example, any of the methods would relate the Cell Type Ontology term *T cell* (CL:0000084) to the two Gene Ontology terms *T cell proliferation* (GO:0042098) and *regulation of T cell proliferation* (GO:0042129), but from an ontological perspective, it would be preferable to link it directly to *T cell proliferation* and then allow interontology linking to connect that concept to *regulation of T cell proliferation*.
- The methods do not differentiate between different relation types.
- The findings suggest methods for predicting which concepts are likely to yield high numbers of false positive matches, e.g. concepts at high positions in the ontology (*cell* (CL:0000000) was a frequent offender) and words that are isomorphic with General English words (such as the *groups* example given above).
- The development of domain-specific tools, such as stemmers, is important.

The stemming/stop word removal technique results deserve special mention. As we noted above, this lexical technique was the least accurate, but the accuracy reported in this study is probably an underestimate.

On a granular examination of the accuracy of each technique in each ontology, it was noted that the results for stemming/stop word removal were particularly abysmal when the query terms were from BRENDA – the accuracy for matching against Gene Ontology definitions was only 15%, and the accuracy for matching against Gene Ontology terms was a stunning 7%. The authors examined the performance of the algorithm on this combination of ontologies and discovered that the low performance noted for BRENDA was the result of a system design bug. Thus, the reported accuracies for stemming and stopword removal for this ontology, as well as the overall average accuracy for stemming and stopword removal in general in these experiments, are almost certainly underestimated. More details on how exactly the design error was discovered are given in the chapter on software testing.

CHAPTER 8

Summarization

Summarization is the task of reducing a document to a shorter form. The two approaches to summarization are *extractive summarization*, in which segments of text are extracted from the original document and assembled into a summary, and *abstractive summarization*, in which novel text is composed to form the summary. Of the two, extractive summarization is far more commonly studied.

Summarization systems may perform *single-document summarization* or *multidocument summarization*. In the former case, only a single document is summarized. In the latter case, the summary is formed from multiple documents.

Summarization systems can also attempt to achieve different compression ratios. The *compression ratio* is the amount by which the length of the original is decreased. Compression ratios result in shorter summaries than compression ratios. Systems may perform better or worse at higher or lower compression ratios.

Summaries themselves can be either indicative or informative. An *indicative* summary lets the user know whether or not they are likely to be interested in reading the full document. An *informative* summary actually communicates some of the information in the document.

8.1 Medical summarization systems

8.1.1 Overview of medical summarization systems

In the medical domain, different approaches to summarization are applied for different tasks, documents (for example, different types of clinical notes in a patient's EHR, MEDLINE citations or full-text documents) and intended users (for example, patients or clinicians[11]).

MiTAP (MITRE Text and Audio Processing) (Damianos *et al.* 2002) (described in Afantenos, Karkaletsis, & Stamatopoulos 2005) is an example of a single-document extractive system. MiTAP monitors epidemiological reports, newswire

11. This subsection draws heavily on the material in Afantenos, Karkaletsis, & Stamatopoulos (2005).

feeds, email, and online news to detect disease outbreaks and other biological threats. Johnson *et al.* (2002) built a multi-document extractive system that takes a user query and a set of documents as input. The query is passed to a search engine, which returns a subset of the documents. The returned documents are clustered and a vector space model is built to represent each cluster. Each sentence of each document is then converted to a vector space model and compared to the cluster vector space model. The highest-ranked sentences are then assembled to form a summary. TRESTLE (Text Retrieval Extraction and Summarisation Technologies for Large Enterprises) (Gaizauskas *et al.* 2001) performs single-document abstractive summarization of pharmaceutical newsletters, using an information extraction approach.

8.1.2 A representative medical summarization system: Centrifuser

A domain- and genre-specific multidocument summarization system, Centrifuser, focuses on patient health care documents (Kan, McKeown, & Klavans 2001b). Centrifuser (for centroid fuser) was developed within the PERSIVAL medical digital library project at Columbia University. In response to a user query, Centrifuser provides both indicative and informative summaries, using topic trees to represent documents and computing tree-based topic similarity. The summarization module extracts salient sentences to summarize documents on similar topics. The relationships between the topics are used to generate navigation links to broader and narrower related topics. The searcher module generates text that describes high level differences between individual topically related documents (Kan, McKeown, & Klavans 2001a).

Centrifuser begins by representing each document in the collection by a hierarchical document topic tree, which breaks each document's topic into subtopics. The topic tree relies on the document structure. For example, the section headers of *The Merck manual of medical information* are used to construct the following first-level subtopics for the topic *Angina*: *Causes*, *Symptoms*, *Diagnosis*, *Prognosis*, *Treatment*. Subtopics beyond document structure are computed using textual features. For example, the *Treatment* subtopic has two subtopics: *Drug Therapy* and *Revascularization Procedures*. The *Drug Therapy* subtopic is further split into several paragraphs each describing a drug or a drug class, for example, *Nitroglycerin* or *Antiplatelet drugs*. To extract these subtopics, Centrifuser processes the text of the paragraphs. Each subtopic in the document tree is enriched with formatting and style metadata. In the absence of rich markup, Centrifuser backs off to recognizing the spatial layout of the document and topic segmentation.

Once each document tree is built, the individual document trees are combined into a composite topic tree for each text type (a set of documents that share the same domain (subject area) and genre, for example, purpose and style). All documents of the same text type are instances of the composite topic tree. The composite topic tree encodes each topic's typicality, length, its position within an article, composition, and variant lexical surface representations of the topic. For example, in the composite topic tree, *Disease*, *Symptoms* (alternatively expressed as *Signs*) is a very typical subtopic that gets a .95 out of 1 score. It usually precedes the sibling subtopics *Treatment* and *Cause* in the document layout, whereas information on prognosis occurs rarely in the consumer health documents. To build the composite topic tree, Centrifuser merges instances of document topic trees of the same type by iteratively aligning most similar subtopics across documents using a similarity metric that combines lexical overlap, subtopic node ordering and depth, and parent topic similarity.

Equipped with a user's query mapped to topic nodes (using string similarity between the query text and the topic's lexical forms) in the document and composite trees, document topic trees for each of the documents in the result set, and a composite topic tree for the text type, Centrifuser generates a multidocument summary. The high-level multidocument synopsis is composed using sentences extracted from each subtopic. The subtopic sentences are first clustered, and then one sentence is selected to represent each cluster using its position in the text, style, length, and other heuristics. The sentences selected for the summary are ordered using subtopic ranking and position in the original text. This stage generates an informative broad summary.

Next, an indicative summary is generated in three stages: content calculation, planning and realization. The potential summary content is computed by determining input topics present in the document set. For each topic, the system assesses its relevance to the query and its prototypicality. Each individual topic in the individual document topic trees is classified as typical, rare, irrelevant or intricate. The distribution of these four types in a document determines its document category as follows: *Prototypical* (contains 50% or more typical topics); *Comprehensive* (contains typical topics and more); *Specialized* (contains some typical topics); *Atypical* (contains 50% or more rare topics); *Deep* (contains 50% or more intricate topics); *Irrelevant* (contains 50% or more irrelevant topics); and *Generic* (contains a mix of topics and exhibits no trends). The text planning phase places the summary document set into the document categories. The indicative summary represents each document category (in the above order). Within each category, the obligatory information (the document category's constituents and its description) is expressed first. Optional information about sample topics or other unusual document features is presented next. Finally, the discourse plan is realized as text. The separate categories are semantically distinct and are realized separately (for example, as separate list items). Two concerns are addressed while generating referring expressions: the size of the document set and applicability of the optional information to all documents in the set. If a category contains more than five documents, the size of the set and an exemplary document title are provided, for example, "There are 23 documents (such as the AMA Guide to Angina) that have detailed information on a particular subtopic of angina". If optional information applies to a subset of documents, the documents are grouped to compress the referring expression, for example, "The first five documents contain figures and tables." The linguistic realization of the sentence plan is rule-based.

Another summarization system developed within the PERSIVAL project produces patient-specific briefings of clinical studies and adapts the summary to the level of expertise of the user (Elhadad *et al.* 2005). A related task-based evaluation showed that for physicians, a patient-specific summary of clinical literature is more useful than a generic summary, and additionally suggested that an abstractive summary is more useful than an extractive summary for this application (Elhadad 2006).

8.2 Genomics summarization systems

In contrast to the medical domain, there has been much less work on summarization in the genomics domain. In general, it has tended to be single-document summarization and to focus on sentence extraction.

The work of Lu *et al.* on automatic GeneRIF production is representative of simple sentence extraction techniques. GeneRIFs are short (less than 255 characters) descriptions of the function (loosely construed) of a gene. They are part of gene entries in the Entrez Gene database. They are typically cut-and-pasted directly from the text of an article abstract. Thus they can be considered to be manually produced single-document extractive summaries.

Although GeneRIFs are useful parts of a gene's Entrez Gene entry, they have always been produced manually, which is a slow process that has been unable to keep up with the rate of genomic data deposition in Entrez Gene. For this reason, Lu *et al.* (Lu, Cohen, & Hunter 2006) tackled the problem of automatic GeneRIF production.

Lu *et al.*'s technique is an example of an application of an early summarization approach known as the Edmundsonian paradigm. Sentences are given points for a generally basic set of features. The sentence with the highest score is returned by the system.

The system iterated over the sentences in an abstract and assigned one point for each of the following features:

- One point is assigned if the sentence is the first, last, or penultimate sentence of the abstract. This feature is based on the observation that sentence position is highly predictive of whether or not a sentence is a good summary sentence.
- One point is assigned if a sentence contains at least one cue word. A *cue word* is a word that indicates that a sentence is likely to be a good abstract sentence. For GeneRIFs, the set of cue words was determined by manual examination and verified by mutual information scores. Table 8.1 gives examples of cue words that were found to be helpful for GeneRIF production.
- One point is assigned for every token that the sentence shares with any Gene Ontology term that has been assigned to that gene. We elaborate on this feature below.

Table 8.1 Cue words for GeneRIFs.
(Supplementary material to Lu, Cohen, & Hunter 2006)

results
novel
review
conclude, concluded
conclusion
role
findings
thus
indicate, indicated, indicates
indication
implicate, implicated, implicates
implication

The Gene Ontology term feature capitalized on an external knowledge source that is available for many Entrez Gene entries. This knowledge source is the set of Gene Ontology terms with which that gene has been annotated (in the model organism database curation sense of the word "annotation" – see the section on user types in Chapter 2). The set of Gene Ontology terms with which the gene has been annotated was tokenized and stemmed. Similarly, each sentence in the abstract was tokenized and stemmed. Stop words were removed. For every Gene Ontology term token that appeared in a sentence, that sentence was assigned one point. Thus, unlike the sentence position and cue word features, which can assign only a single point, the Gene Ontology term feature could result in the assignment of multiple points – this feature was heavily weighted.

Table 8.2 Introductory phrases removed to better match GeneRIF format. (Supplementary material to Lu, Cohen, & Hunter 2006)

We conclude that
These data indicate that
These observations are consistent with the conclusion that
We propose that
From these findings we conclude that
These findings indicate that
This finding suggests that
These results indicate
We show that
These data indicate that
These data suggest that
Our findings support the notion that
Altogether, the results raise the possibility that
Our results suggest that
We speculate that
These results demonstrate that
In conclusion, the results demonstrate that
It is possible that
These observations suggest that
Therefore, it is proposed that
We have thus shown that
Results indicate that
These observations provide strong support for the conclusion that

Finally, the sentence with the highest score was returned as the candidate GeneRIF for that abstract.

The method was evaluated against a gold-standard data set of 413 GeneRIFs from human genes (the method is actually organism-independent and has been applied to three other organisms) that each were associated with genes that had at least one GeneRIF and at least one Gene Ontology term assigned to it that shared the PubMed ID of at least one of the GeneRIFs associated with the gene. The baseline consisted of picking the title of the paper; previous work on automated GeneRIF production had shown that no system was able to perform much better than this baseline. In contrast, Lu *et al.*'s system scores up to 21.3% higher than the baseline (scores were evaluated at different Dice coefficients between the system output and the actual GeneRIF, yielding a number of different scores). The study showed that general summarization techniques such as sentence position and cue words were applicable to highly specialized biomedical text, and that external knowledge sources could be exploited to increase performance of even a simple summarization system. In further experiments, the authors also compared their results to a machine learning system based on the same features and found that the machine learning approach obtained no improvement in results over the simple rule-based approach.

It should be pointed out that Lu *et al.*'s original approach suffers from a scaling problem, since it can only be applied to genes which have Gene Ontology terms associated with them. However, there is no reason to think that the algorithm could not be adjusted to find arbitrary Gene Ontology tokens in the abstract sentences and perform just as well.

8.2.1 Sentence selection for protein–protein interactions

The BioCreative II shared task included a subtask called the ISS, or Interaction Sentence Selection, task. The input was a set of articles. The output was to be the best sentence providing evidence for every protein–protein interaction mentioned in the article. The gold standard was the actual sentences picked from each article for inclusion in the MINT and IntAct databases as evidence for individual protein–protein interactions. The available training data was limited to 63 sentences, so a number of additional test suites were provided. These included a set of 697 positive sentences from Anne-Lise Veuthey, a set of 921 positive and negative sentences from the Prodisen corpus, a set of positive sentences categorized as to whether the interaction described was direct or indirect from Christine Brun, and a set of 51,381 GeneRIFs, all positive examples. They all differed from the actual test set in that they were drawn from abstracts, while the test data consisted of full-text articles, and in that they were single sentences, while gold-standard "sentences" could comprise several sentences. The test set consisted of 358 full-text articles. The most successful approach was an adaptation of the Lu *et al.* approach for GeneRIF prediction described above. In a first step, all sentences were filtered, with the filtering differing somewhat for different parts of the article. Screening features included the presence of positive cue words, the absense of negative cue words (stigma terms), having at least one gene mention, containing a mention of an experimental method, and having an interaction word.

Any sentence that passed that filtering step were assigned a score, with the highest-scoring sentence then being returned. Scores were assigned as follows:

- Any mention of an experimental method gets ten points.
- Any interaction key word or positive cue term gets one point for each token in the key word or cue term.
- High-frequency words from the gold standard sentences receive a point.
- Sentences in the results, title, abstract, or introduction sections receive a point.
- Mentions of gene/protein names receive a point.
- Mentions of figures or tables receive a point.

Other methods were applied by other teams, particularly machine learning methods. In general, systems looked for multiple protein names, mention of experimental methods, and sentences in specific sections of the article, including headers of sections and subsections and figure legends. Some systems also made use of syntactic parse trees, and others made use of word weightings derived from the training sentences (Krallinger, personal communication). However, the Edmundsonian paradigm was difficult to beat.

8.2.2 EntrezGene SUMMARY field generation

Another important potential real-world application of multi-document extractive summarization is the generation of SUMMARY fields for EntrezGene entries. SUMMARY fields (see also the chapter on information retrieval) contain a small set of sentences that summarize what is known about a gene in a semi-structured form. At the time of writing, there are about 5,000,000 genes in PubMed, and only about 20,000 have a SUMMARY field. Jin *et al.* (2009) describe a system for automatically generating SUMMARY fields. Documents about a gene of interest are retrieved and their sentences are filtered for informativeness based on the number of words that they contain that have been empirically determined from manual summaries to be indicative of summary-worthiness, known as topic signature terms. Table 8.3 gives examples of unigram topic signature terms. Any sentences that pass the filtering process are scored based on a lexical PageRank algorithm and the similarity between the sentence and the Gene Ontology terms with which that gene has been annotated (in the biological sense). The system was evaluated against a set of 7,294 manually generated SUMMARY fields and achieved a ROUGE-1 score of 0.4725, ROUGE-2 of 0.1247, and ROUGE-SU4 of 0.1828. (ROUGE is a family of metrics for evaluating summarization systems, based on string similarity between a system output and a gold standard, calculated with varying levels of stringency and definitions of "string"). This outperformed both the MEAD system and a random baseline.

Table 8.3 Unigram topic signature terms for SUMMARY field sentence selection from Jin *et al.* (2009).

protein	*member*	*receptor*
gene	*variant*	*isoform*
encode	*domain*	*alternative*
family	*splice*	*bind*
transcription	*subunit*	*involve*

CHAPTER 9

Question-answering

Question answering, the process of generating answers to questions in natural language, is one of the basic human activities. People generally satisfy their information needs and learn by asking questions, and ultimately would like to have systems capable of understanding their questions and generating adequate answers. The currently available question answering systems approximate the solutions to this practical problem through software engineering approaches that draw on tools and understanding of the question answering process developed in psychology, medicine, philosophy, linguistics, education, and computer and library sciences.

9.1 Principles

Schematically, there are four parts to automatic question answering: (1) understanding the meaning (type) of a question and the underlying information need; (2) representing the question in the form compatible with the knowledge base of the system; (3) finding candidate answers; and (4) generating a concise, yet complete, answer along with the answer support (information about the sources of the answer and their authoritativeness).

9.1.1 Question analysis and formal representation

To date, analysis of clinical questions and information needs of clinicians is much better researched than information needs of other biomedical researchers and practitioners.

9.1.1.1 *Clinical questions*

Information needs of health professionals have been actively researched for over two decades. The early studies focused on the types of information needs clinicians have and the number of questions arising due to attending to patients, as well as on the types and proportion of needs that are being pursued and on the most likely sources of information. One of the early reviews of information needs research (Smith 1996) categorized information needs of health professionals as follows:

- information on particular patients,
- data on health and sickness within the local population,
- medical knowledge,
- local information on doctors available for referral,
- information on local social influences and expectations, and
- information on scientific, political, legal, social, management, and ethical changes affecting both how medicine is practiced and how doctors interact with individual patients.

Of these, answering medical knowledge questions was until recently the only focus of NLP research. Current work on providing information about particular patients within electronic health records and on health and sickness within the local population (the focus of epidemiology and a relatively recently introduced term biosurveillance) is discussed in Demner-Fushman, Chapman, & McDonald (2009).

Clinical questions occur due to lack of medical background and foreground knowledge (Richardson & Wilson 1997). The background knowledge is a general knowledge of the basic facts of a disease. Information needs with respect to the background knowledge will generate information requests in the form of wh-type questions (who, what, where, when, why, and how) such as *What is this disorder?*, *What causes it?*, and *How should I treat this disorder?*

Foreground questions are more specific and "patient-centered". These complex questions address the components of a problem, the details of interventions, and the potential clinical outcomes in a specific clinical situation. More often, practicing clinicians lack the foreground knowledge, for example, about diagnostic tests best suited for a particular patient, or the best currently recommended treatment strategy for a given condition.

The foreground information needs of clinicians were summarized by Ely *et al.* (2005). The 48 physicians observed by Ely *et al.* reported that their needs would be satisfied by comprehensive resources that answered their questions and rapid access to concise answers (comprehensive specific bottom-line recommendations) that were easy to find and told them what to do in highly specific terms, providing evidence-based rationale for recommendations.

9.1.2 Formal representation of questions

There are two approaches to modeling and formally representing clinical questions: empirical and based on domain models. In the empirical approaches, the questions asked by clinicians at the point of care are collected and then systematized into a limited number of generic question types. For example, a taxonomy derived using over 1,300 collected questions consists of four hierarchical levels (Ely *et al.* 2000).

The top level of this taxonomy contains five broad areas: diagnosis, treatment, management, epidemiology, and non-clinical questions. The lower levels focus on the aspects of the broad areas, resulting in 64 quaternary categories. For example, 24 original questions were reduced to the generic quaternary category "Is drug x safe to use in situation y? OR Is drug x contraindicated in situation y?" that covers the following aspects: treatment (primary), drug prescribing (secondary), adverse effects (tertiary), and safety contraindications (quaternary).

In early explorations of generic question types, clinical questions were formally represented as concept-relation-concept triplets expressed with nineteen Semantic Network relations and forty UMLS semantic types (for example, [pharmacologic substance]-(treats)-[diseases or syndrome]) (Cimino *et al.* 1993). The triplet-based question representation ([concept]-(relation)-[concept]) can also be used to find answers on the World Wide Web (Jacquemart & Zweigenbaum 2003).

The triplets could also be used to formally represent questions in a question answering system that uses a relational database as its knowledge base. For example, the database could contain a drug treatment indications relation, and the relation attributes would include disease names and drug names. To answer questions about drugs indicated for a given disease and diseases for which a given drug is indicated, the system would represent the questions as SQL queries. The relation tuples retrieved by the query could be presented as answers. Alternatively, if the relations were extracted from a scientific publication, the text segments corresponding to the tuples could be returned as the answer (Katz, Lin, & Felshin 2001).

9.1.3 Domain model-based question representation

Domain models generally define, on a high level, the main entities and actors in the domain, and relations between the entities.

One of the domain models was developed in the context of evidence-based medicine (EBM). EBM is a widely accepted paradigm that formalizes approaches to bridging the gap between the care that a patient will get and the best possible care in a given situation as determined by systematic research (Sackett *et al.* 2000). The conceptual model developed within EBM consists of three major components: (1) a framework that captures the elements of a clinical scenario, (2) clinical tasks that encompass the actors and relations between the elements of a clinical scenario, and (3) the strength of clinical research evidence. The framework for capturing clinical scenarios consists of four slots: P – a slot that defines the patient characteristics and the problem that brought the patient to the clinician; I and C that define the intended intervention and a comparison; and O – the desirable patient outcome. The relations between the slots are determined by the clinical task. For example, if the task is treatment, the intervention is a therapeutic procedure or a drug that treats the problem, but if the task is diagnosis, the intervention is a diagnostic test that differentiates between problems with similar symptoms.

Incorporating the best available research evidence in decision making, much like question answering, involves: defining the clinical question; finding the best information; appraising the information for validity and relevance; and summarizing the information (Rosenberg & Donald 1995). Specific guidelines for formulating clinical questions and finding and appraising evidence were published in a series of articles in the Journal of the American Medical Association (JAMA) (Guyatt, Sackett, & Cook 1994; Jaeschke, Guyatt, & Sackett 1994; Levine *et al.* 1994; Laupacis *et al.* 1994). These guidelines roughly correspond to reference interviews conducted by reference librarians and can be used in automatic question answering (Booth & O'Rourke 1997).

Questions occur while clinicians perform all clinical tasks. The Task Force rating scheme (AHRQ 2002) for assessing the quality of clinical evidence identifies the following major clinical tasks:

Etiology/Harm identifying causes for diseases or conditions

Diagnosis encompasses clinical findings, diagnostic tests, and differential diagnosis

- **Differential diagnosis** identifying and estimating likelihood of potential causes for patient's condition
- **Diagnostic test** selecting and interpreting diagnostic tests, considering their precision, accuracy, acceptability, cost, and safety

Therapy select treatments that are worth the efforts and costs of using them (includes **Prevention** – actions to reduce the chance of disease by identifying and modifying risk factors)

Prognosis estimate the patient's likely course with time and anticipate likely complications

EBM provides a task-independent framework for formulating clinical questions that consists of four elements:

P Who/What is the **P**atient/**P**roblem being addressed?
I What is the intended **I**ntervention?
C What is the intervention **C**ompared to?
O What are the **O**utcomes?

Questions containing the EBM-based clinical question structure elements (known by the first letters of the elements of the question frame as *PICO*) are thought to be "answerable" (Richardson *et al.* 1995).

Other domain models augment the *PICO* structure, for example, with the number of subjects in a research study and statistics used to analyze the study (Flaherty 2004). Yet other models encompass the declarative ("what to know") and procedural ("what to do") knowledge needed for clinical problem solving. For example, Florance (1992) identified four states (characterized in frame-like structures) that captured information needs arising in the decision-making processes (prediagnostic assessment, diagnosis, treatment choice, and learning). Each clinical problem solving frame consisted of slots and each slot had one or more facets. For example, the prediagnostic assessment frame had four slots: condition, evidence, treatment, and processes. The prediagnostic assessment condition slot was, in turn, comprised of four facets: status (for example, "chronic"), features (for example, "cause"), relationships (for example, "co-occurrence"), and prognosis (for example, "cure").

9.1.3.1 *Genomics and translational research questions*

Genomics and translational research questions were studied in the context of the Genomics track of the Text REtrieval Conference (TREC). Starting in 2003, track participants interviewed biologists to collect their information needs. In 2005, the track steering committee used the biologists' information needs to generate five Generic Topic Types. The topic types (later validated in collection of new information needs in interviews with biologists) included information about:

- standard experimental methods or protocols,
- role of a gene in a given disease,
- role of a gene in a specific biological process,
- interactions (for example, promote, suppress, inhibit) between two or more genes in the function of an organ or in a disease,
- mutations of a given gene and its biological impact

Subsequently, the generic templates were expressed using four question patterns:

- What is the role of gene in disease?
- What effect does gene have on biological process?
- How do genes interact in organ function?
- How does a mutation in gene influence biological process?

Later, biologists' information needs were represented as question types based on the types of named entities constituting answers (ANTIBODIES, BIOLOGICAL SUBSTANCES, CELL OR TISSUE TYPES, MOLECULAR FUNCTIONS, PATHWAYS, etc.). For example, "What [TUMOR TYPES] are associated with Rb1 mutations?" (Hersh & Voorhees 2009).

9.1.4 Answer retrieval

A system that stores answer frames (for example, *associated(GENE,TUMOR)* or *PICO*) would retrieve answers directly through matching the questions and the answers. Alternatively, if relatively short text passages (for example, paragraphs) are considered appropriate answers (as was the case in the Genomics track), a search engine that indexes paragraphs would return the top ranking passages as answers.

In practice, the existing systems resort to retrieving a set of potentially topically relevant documents using conventional search engines, and post-process retrieval results. The post-processing steps range from reducing the size of retrieved passages to filtering and re-ranking retrieval results and answer extraction.

Using a search engine to reduce the set of documents to those that potentially contain answers imposes an additional query formulation step. The specifics of this step are dictated by the search engine. For example, if PubMed is used to answer one of the most frequently occurring questions: "What is the best drug treatment for X?", the search template might be "X/dt[majr]", where *X* would be replaced with the problem name, *dt* stands for the *drug therapy* subheading, and *majr* indicates that only papers focused on drug therapy are of interest. For example, submitting the question "What is the best drug treatment for pneumonia?" as "pneumonia/dt[majr]", results in query translation "pneumonia/drug therapy[MAJR]", and only papers containing corresponding Mesh Terms (for example, "Pneumonia/*diagnosis/*drug therapy/microbiology/therapy") will be retrieved. Such queries are usually specific, but lack sensitivity. For greater recall in this step, a system might submit the whole question (with or without removing function words). In PubMed/MEDLINE's case, this will result in the following query translation:

> best[All Fields] AND drug[All Fields] AND (therapy[Subheading] OR therapy[All Fields] OR treatment[All Fields] OR therapeutics[MeSH Terms] OR therapeutics[All Fields]) AND (pneumonia[MeSH Terms] OR pneumonia[All Fields])

Clearly, the post-processing steps for the second query formulation strategy will have to account for potentially marginally related documents, unlikely to contain answers. A system might apply additional filters before or after the answer extraction step.

9.1.5 Answer extraction and generation

Ideally, the answers generated by a QA system will follow the answer format expected by the users. The format for genomics and translational research questions developed in the Genomics track question answering task consists of text passages automatically extracted from scientific publications and judged relevant by domain experts and named entities manually extracted from the relevant passages by the domain experts.

For example, *C-KIT GENE* was the entity extracted from three paragraphs that answer the question "What [GENES] are involved in the melanogenesis of human lung cancers?" One of the passages supporting this answer:

> The proto-oncogene c-kit encodes a transmembrane tyrosine kinase receptor (c-KIT/CD117) related to the platelet-derived growth factor (PDGF)/colony-stimulating factor 1 (CSF-1) (c-fms) receptor subfamily. CD117 is thought to play an important role in hematopoiesis, spermatogenesis, melanogenesis and, more recently, in carcinogenesis.Overexpression of CD117 has previously been documented in myeloid leukemia, neuroblastoma, breast tumor, colon tumors, gynecological tumors, testicular germ cell tumors and SCLC.

Several manually curated clinical question answering sources provide examples of the formats of answers to clinical questions.

9.1.5.1 *Reference answer formats for clinical questions*

Some clinical question answering sources provide answers in short passages. For example, Parkhurst Exchange (2010) gives the following answer to the question: "What's the updated treatment for community-acquired pneumonia (CAP)?":

> This is not a simple question to answer. Although CAP is often referred to as a single entity, it's actually heterogenous and ranges from mild to life-threatening disease. A number of features can influence why and how it develops. When patients with extensive healthcare exposure and comorbid illnesses develop CAP, they're at greater risk for resistant organisms. On the other hand, CAP that arises as a result of aspiration will require the addition of anti-anaerobic agents. Immunocompromised individuals will have yet a different range of infective causes. Finally, outbreaks can stem from a variety of etiologies – e.g. influenza virus, methicillin-resistant Staphylococcus aureus or pneumococcus in the homeless or injection drug users. The root cause at hand will have a major influence on the choice of therapy.

Note that the answer would have been shorter if the question contained more details about the patient. Other sources answer more focused questions and use a hierarchical answer format that provides a short bottom line and then expands it in increasingly detailed layers.

For example, BMJ Clinical Evidence (2010) first answers a similar question, "What are the effects of treatments for community-acquired pneumonia in people admitted to hospital?" with a ranked list of treatments:

- Likely to be beneficial
  - Antibiotics in hospital (compared with no antibiotics)
  - Early mobilisation (may reduce hospital stay compared with usual care)
- Unlikely to be beneficial
  - Intravenous antibiotics in immunocompetent people in hospital without life-threatening illness (compared with oral antibiotics)

Then the answer summarizes the results of specific studies, in one short paragraph each (the example answer contains four paragraphs for four specific studies that compared the following antibiotics: clarithromycin and erythromycin; sparfloxacin and clarithromycin; azithromycin and levofloxacin; and azithromycin compared with clarithromycin). Then the answer discusses the benefits and harms of the treatments in details, and provides references to the original studies.

Similarly, the answers to Clinical Inquiries in the Journal of Family Practice (2010) present short, one-sentence fast track answers, and more detailed evidence-based answers, evidence summaries discussing individual studies used to generate the answers, bottom-line recommendations and references. For example, the short and the evidence-based answers to the question "Can community-acquired pneumonia be treated with 3 to 5 days of antibiotic therapy?" are:

> A 3- to 5-day course of antibiotics treats community-acquired pneumonia just as well as longer courses
>
> Yes. A 10- to 14-day course of antibiotics is no more effective for patients with community-acquired pneumonia than 3 to 5 days of treatment. Clinical failures and mortality were similar regardless of treatment length. This study demonstrated equivalent effectiveness for oral or parenteral azithromycin for 3 to 5 days, levofloxacin for 5 days, cefuroxime for 7 days, and intravenous ceftriaxone for 5 days.

9.1.5.2 *Entity-extraction approaches to answer generation*

In the entity-extraction approach, the documents are searched for entities found in the question and entities of the answer type. Then the sentences or passages containing the terms are extracted and ranked by density of the relevant entities. For example, to answer the question "What [TUMOR TYPES] are associated with Rb1 mutations", a system could search for "Rb1 mutations AND neoplasms[majr]" and then extract specific entities having the semantic type "Neoplastic Process" using co-occurrence of the search term "Rb1 mutations" and an entity in a sentence, or establishing an "associated with" relation between the terms, which could be accomplished for the following titles (returned by the search among others):

- "Detection of mosaic RB1 mutations in families with retinoblastoma"
- "Constitutional and somatic RB1 mutation spectrum in nonfamilial unilateral and bilateral retinoblastoma"
- "Identification of ARHGEF17, DENND2D, FGFR3, and RB1 mutations in melanoma"

The danger of using simple co-occurrence of entities is demonstrated in the following passages:

- "a gene(s) telomeric to RB1 is involved in the malignant transformation of CLL"[Chronic Lymphocytic Leukemia]
- "inactivation of both RB1 alleles in mouse urothelium failed to accelerate urothelial proliferation"

More sophisticated processing will establish that in the first passage RB1 is used to localize the involved genes, and the role of RB1 in urothelial tumorigenesis is negated in the second passage.

9.2 Applications

In this section we will describe the CQA-1.0 system, based on the semantic domain model (Demner-Fushman & Lin 2007). The system supports several NLM applications as follows: given (or inferring) a user's question, the CQA-1.0 system formulates a query, finds documents using the Essie search engine (Ide, Loane, & Demner-Fushman 2007b), estimates their validity and relevance to the question for each document, discards potentially irrelevant documents, re-ranks relevant documents, extracts answers, and presents information in a multi-tiered answer.

The steps automatically performed by the CQA-1.0 system are based on three fundamental components identified in the EBM-based semantic domain model: clinical task, framework for question formulation and document appraisal, and strength of evidence. The architecture of the system is shown in Figure 9.1.

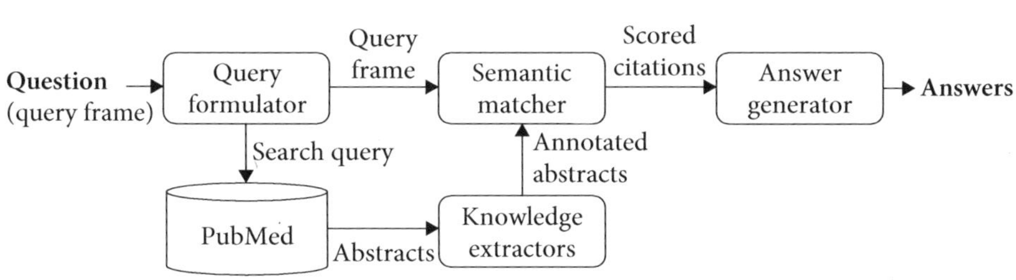

Figure 9.1 Overall architecture of the CQA-1.0 system.

The clinical question structure known by the first letters of the elements of the question frame as PICO plays a role not only in the question and search formulation, but also in appraisal of the information. Although the elements of the clinical scenario encoded in the PICO frame are constant for all clinical tasks, the semantic types that populate the frame are determined by the task. For example, if the task is diagnosis, the intervention slot of the frame will be populated by diagnostic procedures. The third important component of the EBM-based semantic domain model is the strength of the found clinical evidence. The strength of evidence in a clinical study is determined based on soundness of its design, number of patients participating in the study, and methods used to evaluate the results of the study.

9.2.1 Question analysis and query formulation

The system provides a PICO form that can be filled out by a user (including clinical task), a simple text input field, and an API for submission of short clinical notes. The problems, patients' characteristics and interventions are extracted from the submitted free text using the knowledge extractors described below. For example, given a progress note for a cancer patient:

> Start a bowel regimen to avoid constipation; Planned Interventions: colace/senna ordered daily, miralax ordered PRN;

The system extracts *constipation* as Problem, and *colace*, *senna*, and *miralax* as Interventions, *regimen* is recognized as a general-purpose *Therapeutic or Preventive Procedure* and the PICO frame is filled as follows:

> *search task:* therapy selection
> *problem:* constipation
> *population:* cancer
> *intervention:* colace, senna, miralax

Given the frame, the query is formulated by ANDing patient's characteristic and problems with ORed interventions: "CANCER AND (Constipation AND (Colace OR Senna OR Miralax))".

An example answer to the question is found in PubMed:

> A comparison of sennosides-based bowel protocols with and without docusate in hospitalized patients with cancer. Thirty patients received the sennosides-only (S) protocol and 30 the sennosides plus docusate (DS) protocol. Over a total of 488 days of observation it was found that the S protocol produced more bowel movements than the DS protocol, and in the symptom control/supportive care patients this difference was statistically significant. The addition of the initial

> docusate-only step and adding docusate 400–600 mg/d to the sennosides did not reduce bowel cramps, and was less effective in inducing laxation than the sennosides-only protocol.

If a query is unsuccessful, the population slot is omitted and the problems are ORed.

9.2.2 Knowledge Extraction

Each knowledge extractor takes as input clinical text or the abstract text of a MEDLINE citation and identifies the relevant elements: the problems and interventions are short noun phrases for any input text. Patients' demographics (age, gender, etc.) and anatomical locations of the problems are extracted from clinical notes, whereas the number of patients in the study, demographics and the supporting sentence are extracted from the abstract. The outcome statements – 2–3 sentences that summarize the results of the study – are extracted as answers from the abstracts.

The knowledge extractors rely extensively on MetaMap (Aronson 2001b) and built-in named entity extraction mechanisms to extract concepts in Semantic Groups (McCray, Burgun, & Bodenreider 2001) Disorders and Drugs, and Semantic Types Therapeutic or Preventive Procedure and Diagnostic Procedure. The extractors use the sections of the structured abstracts. For example, information about study population is mostly found in the methods section.

The population, problem, and the intervention extractors are based largely on recognition of semantic types and a few manually constructed rules; the outcome extractor is implemented as an ensemble of classifiers trained using supervised machine learning.

9.2.2.1 *Population Extractor*

Population elements consist of the number of patients in the study, the population group (for example, adolescents) and the sentence that contains the identified population.

- The concept describing the population belongs to the semantic type Group or any of its children. In addition, certain nouns are often used to describe study participants in medical texts; for example, an often observed pattern is "subjects" or "cases" followed by a concept from the semantic group Disorder.
- The number of subjects that participated in the study often precedes or follows a concept identified as a Group. In the latter case, the number is sometimes given in parentheses using a common pattern n=*number*, where "n=" is a shorthand for the number of subjects, and *number* provides the actual number of study participants.

- The confidence that a clause with an identified number and GROUP contains information about the population is inversely proportional to the distance between the two entities.
- The confidence that a clause contains the population is influenced by the position of the clause, with respect to headings in the case of structured abstracts and with respect to the beginning of the abstract in the case of unstructured abstracts.

Given the above assumptions, the population extractor searches for the following patterns:

- GROUP ([Nn]=[0–9]+)
  for example, "the elite endurance athletes (n=410)"
- *number* + DISORDER? GROUP
  for example, "Thirty patients, 265 prostate cancer patients"

The confidence score assigned to a particular pattern match is a function of both its position in the abstract and its position in the clause from which it was extracted. If a number is followed by a measure – for example, *year* or *percent* – the number is discarded, and pattern matching continues. After the entire abstract is processed in this manner, the match with the highest confidence value is retained as the population description.

9.2.2.2 *Problem Extractor*

The problem extractor relies on the recognition of concepts belonging to the UMLS semantic group Disorder. The confidence score given to a concept is a function of its location in the discourse and frequency. Concepts in the title, in the introduction section of structured abstracts, or in the first two sentences in unstructured abstracts, and in MeSH (if available) are given higher confidence values due to their discourse prominence. Finally, the highest-scoring problems are designated as primary problems in order to differentiate them from co-occurring conditions identified in the abstract.

9.2.2.3 *Intervention Extractor*

The intervention extractor identifies both the intervention and comparison elements in a PICO frame; processing of these two frame elements can be collapsed together because they belong to the same semantic group. Restrictions on the semantic types allowed in the UMLS Semantic Network relations associated with each clinical task prescribe the set of possible interventions. At present, the system identifies the following intervention subgroups: DIAGNOSTIC PROCEDURE, PHARMACOTHERAPY, PREVENTIVE PROCEDURE, and THERAPEUTIC PROCEDURE.

Similarly to problems, higher confidence scores are given to concepts of the relevant semantic type if they appear in the title, aims, and methods sections of a structured abstract. In unstructured abstracts, concepts towards the beginning of the abstract text are favored. Finally, the intervention extractor takes into account the presence of certain cue phrases that describe the aim and/or methods of the study, such as "This study examines" or "This paper describes".

9.2.2.4 *Outcome Extractor*

The outcome extractor assigns a probability of being an outcome to each sentence in an abstract. The probability is assigned by an ensemble of classifiers that includes: a rule-based classifier, a unigram "bag of words" classifier, an *n*-gram classifier, a position classifier, an abstract length classifier, and a semantic classifier. With the exception of the rule-based classifier, all classifiers were trained on the 275 citations manually annotated by clinicians trained in medical informatics.

Knowledge for the rule-based classifier was hand-coded, prior to the annotation effort, by a registered nurse with 20 years of clinical experience. This classifier estimates the likelihood that a sentence states an outcome based on cue phrases such as "significantly greater", "well tolerated", and "adverse events". The likelihood of a sentence being an outcome as indicated by cue phrases is the ratio of the cumulative score for recognized phrases to the maximum possible score. For example, the sentence "The dropout rate due to adverse events was 12.4% in the moxonidine and 9.8% in the nitrendipine group" is segmented into eight phrases by MetaMap, which sets the maximum score to 8. The two phrases *dropout rate* and *adverse events* contribute one point each to the cumulative score, which results in likelihood estimate of 0.25 for this sentence.

The unigram "bag of words" classifier is a Naïve Bayes classifier that outputs the probability of a class assignment.

The *n*-gram based classifier is also a Naïve Bayes classifier, but it operates on the most informative unigrams and bigrams identified using the information gain measure, and then reduced to the positive outcome predictors using odds ratio. This classifier also outputs the probability of a class assignment.

The position classifier returns the maximum likelihood estimate that a sentence is an outcome based on its position in the abstract (for structured abstracts, with respect to the results or conclusions sections; for unstructured abstracts, with respect to the end of the abstract).

The abstract length classifier returns a smoothed (add one smoothing) probability that an abstract of a given length (in the number of sentences) contains an outcome statement. For example, the probability that a four-sentence-long abstract contains an outcome statement is 0.25, and the probability of finding an outcome in a ten-sentence-long abstract is 0.92. This feature turns out to be useful because the average length of abstracts with and without outcome statements differs: 11.7 sentences for the former, 7.95 sentences for the latter.

The semantic classifier assigns to a sentence an *ad hoc* score based on the presence of UMLS concepts belonging to semantic groups highly associated with outcomes such as THERAPEUTIC PROCEDURE or PHARMACOLOGICAL SUBSTANCE. The score is given a boost if the concept has already been identified as the primary problem or an intervention.

The probabilities and likelihood estimates of being an outcome statement (assigned to a sentence by the base classifiers and used as probabilities) are then combined by the meta-classifier using stacking – a version of least squares linear regression adapted for classification (Ting & Witten 1999). This multiple linear regression (MLR) meta-classifier is described by the following equation:

$$LR(x) = \sum_{k=1}^{N} \alpha_k P_k(x) \tag{9.1}$$

$P_k(x)$ is the probability that sentence x belongs to an outcome statement, as determined by classifier k. To predict the class of a sentence, the probabilities generated by K classifiers are combined using the coefficients ($\alpha_0, \ldots, \alpha_k$). The coefficients' values are determined in the training stage as follows: probabilities predicted by base classifiers for each sentence are represented as a $K \times N$ matrix A, where N is the number of sentences in the training set, and K is the number of classifiers. The reference set class assignments for each sentence are stored in a vector b, and the coefficients' values are found by calculating the vector α that minimizes $||A\alpha - b||$. The coefficients were found using singular value decomposition (SVD), as provided in the JAMA basic linear algebra package released by NIST[12].

The knowledge extraction process results in filling the Problem, Population, Intervention and Outcome slots of the document frame. Each slot contains the surface representation of the element(s), the UMLS identifiers (if applicable), and the score that reflects the CQA-1.0 confidence in the identified element.

9.2.2.5 *Clinical Task classification*

The identification of the elements of a clinical scenario (PICO knowledge extraction) is followed in the semantic processing by the identification of the clinical task under study. To determine the task-specific orientation of a MEDLINE citation, it is processed using six binary rule-based task classifiers: therapy, prevention, diagnostic methods, differential diagnosis, etiology, and prognosis. Each classifier returns a confidence score that a study described in the article focused on a given

12. http://math.nist.gov/javanumerics/jama/

clinical task. The score for each clinical task is based on the terms that are positive and negative indicators. Positive indicators for each task were derived from: (1) the PubMed Clinical Query filters (Haynes *et al.* 1994; Wilczynski, McKibbon, & Haynes 2001); (2) the JAMA EBM tutorial series on critical appraisal of medical literature; (3) MeSH scope notes; and (4) observations. A set of positive indicators for the non-clinical orientation is used as an additional set of negative indicators common to all tasks. The terms *genetics* and *cell physiology*, for example, were originally developed as positive indicators for genomics and other basic scientific research articles. The positive and negative weights assigned to each term heuristically encode the relative importance of different MeSH headings. The task score, S_{task}, is given by:

$$S_{\text{task}} = \sum_{t \in \text{MeSH}} \alpha(t) \qquad (9.2)$$

The function $\alpha(t)$ maps a MeSH term to a positive score if the term is a positive indicator for that particular task type, or a negative score if the term is a negative indicator for the clinical task. The highest S_{task} score determines the primary orientation of the study described in the article. Since the classifiers rely on MeSH headings assigned by indexers based on the full text of an article, it is appropriate to assume the task classifiers determine the orientation of the whole article, and not just that of the abstract. Although at present the classifiers use only MeSH headings assigned manually by indexers, should a need arise, the system could rely entirely upon automatic Medical Text Indexing that is currently suggesting terms for indexers' review (Aronson *et al.* 2004). This might lead to approximately 20% degradation in performance.

Indicators and score for therapy task

The examples of strong positive therapy indicators derived from the Clinical Query filters, MeSH scope notes, and JAMA EBM tutorials are: *treatment outcome*, *drug combinations*, *drug therapy*, *therapeutic use*, *surgery*, and *radiotherapy*. A score of 1 is given if the above MeSH descriptor or qualifier is marked as the main theme of the article (indicated via the star notation by indexers), and a score of 0.5 otherwise. The starred non-clinical indicators decrease the score by 1, and by 0.5 otherwise. In addition, two MeSH sub-trees were observed to be weak negative indicators of the task (with a score decrement of 0.1), and one sub-tree as a weak positive indicator (with a score increment = 0.2). The weak positive indicators are *drug administration routes* and any of its children in the MeSH hierarchy. The weak negative indicators are *Health Care Economics and Organizations* and *Health Services Administration*.

Indicators and score for prevention task
The following MeSH terms are considered positive indicators and each add 0.8 to the score:

MeSH Qualifiers:	preventive medicine, primary prevention, life style, risk, risk factors, health behavior, infection control, epidemiologic methods.
MeSH Descriptors:	prevention & control, epidemiology, prevention, prophylaxis, preventive therapy, preventive measures, control.

Indicators and score for differential diagnosis
A single strong positive indicator of differential diagnosis is the term *differential diagnosis*. The remaining diagnosis MeSH terms such as *diagnosis*, *diagnostic use*, *findings*, *examination*, *diagnostic tests*, *predictive value of tests*, *sensitivity and specificity*, etc. are weak positive indicators. The non-clinical and therapy indicators are weak negative indicators for this task (with score decrements 0.4 and 0.1).

Indicators and score for diagnostic methods
Positive indicators for therapy are also used as negative indicators for diagnosis because the relevant studies are usually disjoint. It is highly unlikely that the same clinical trial will study both diagnostic methods and treatment methods. The MeSH term *diagnosis* and any of its children are considered positive indicators. As with therapy questions, MeSH terms marked as the major theme get a score of ±1.0, and ±0.5 otherwise. To distinguish clinically oriented *diagnostic methods* from the research oriented *Investigative Techniques*, this term and all its children, for example, *Animal Experimentation*, are used as weak negative indicators, decreasing the score by 0.5. This rule might be changed or, resources permitting, learned in the future when clinical implications of methods in this sub-tree, for example, *Cytogenetic Analysis* will be of practical interest.

Indicators and score for prognosis task
Positive indicators for prognosis include the following MeSH terms: *survival analysis*, *disease-free survival*, *treatment outcome*, *health status*, *prevalence*, *risk factors*, *disability evaluation*, *quality of life*, and *recovery of function*. For terms marked as the major theme, a score of +2 is given; +1 otherwise. There are no negative indicators (other than those common to all tasks).

Indicators and score for etiology task
Negative indicators for etiology include strong therapy-oriented MeSH terms; these terms are given a score of –0.3. Positive indicators for diagnostic methods and differential diagnosis are weak positive indicators for etiology, and receive a

positive score of 0.1. The following MeSH terms are considered highly indicative of citations relevant to etiology: *population at risk*, *risk factors*, *etiology*, *causality*, and *physiopathology*. If one of these terms is marked as the major theme, a score of +2 is given; otherwise, a score of +1 is given.

9.2.2.6 *Strength of Evidence classification*

The semantic processing of a MEDLINE citation concludes with the identification and scoring of the third basic element of the EBM-based semantic domain model – the Strength of Evidence. The Strength of Evidence indicates how influential a given MEDLINE citation should be in contributing to a clinical decision. Several factors determine the Strength of Evidence: (1) the type of the clinical study, (2) the authority and orientation of the journal in which the article was published, and (3) the recency of the publication. Given these factors, the Strength of Evidence score of a citation is determined as a sum of the scores for each factor:

$$S_{SoE} = S_{study} + S_{journal} + S_{date} \tag{9.3}$$

Strength of the study

Metadata associated with most MEDLINE citations are extensively used in determining strength of evidence, and scoring of its three components. The first component is the type of a clinical study. The potential highest level of the strength of evidence for a given clinical study type can be identified using the Publication Type and MeSH terms pertaining to the type of the clinical study assigned by indexers.

Table 9.1 shows the publication type and MeSH terms mapped to evidence grades according to the principles defined in the Strength of Recommendations Taxonomy (Ebell *et al.* 2004).

Table 9.1 Publication Type and MeSH-based strength of evidence categories.

Strength of Evidence	Publication Type/MeSH
Level A(1)	Meta-Analysis, Controlled Clinical Trials, Randomized Controlled Trials, Multicenter Studies, Double-Blind Method, Cohort Studies, Follow-up Studies
Level B(2)	Studies: Case-Control, Cross-Sectional, Cross-Over, Evaluation, Longitudinal, Retrospective, Case Series
Level C(3)	Case Report, In Vitro, Animal and Animal Testing, Alternatives studies

Level A publication types and MeSH terms increase the overall study type score by 0.5; Level B, by 0.3; Level C by 0.2. The highest evidence level is used to score citations with several publication types and MeSH terms pertaining to different evidence levels. All non-clinical publications decrease the score by 2. Otherwise, a zero score is assigned to S_{study}.

Journal contribution to the score
Citations published in core and high-impact journals such as the Journal of the American Medical Association (JAMA) get a score of 0.6 for $S_{journal}$. The score increases by 0.3 for citations published in one of the approximately 100 journals most likely to contain patient oriented outcomes (identified by the group of clinicians that developed the Strength of Recommendations Taxonomy), for example, in the American Family Physician journal. The remaining journals get a zero score.

Recency of the study
Finally, recency contributes to the strength of evidence score according to Equation 9.4.

$$S_{date} = (year_{publication} - year_{current})/100 \tag{9.4}$$

A mild penalty decreases the score of a citation proportionally to the time difference between the date of the search and the date of publication.

The assignment of the Strength of Evidence score concludes the semantic processing.

9.2.2.7 *Document scoring and ranking*

The semantic processing described above results in a set of document frames fully annotated with the semantic domain model elements. The document frames contain: (1) the elements of a clinical scenario (PICO), (2) a set of confidence scores for each clinical task, and (3) a score for the strength of evidence of the study. The document set is now ready to be ranked with respect to its relevance to the question. The ranking takes place in the CQA-1.0 Semantic Matcher module.

Formally, the relevance of a citation with respect to a clinical question includes contributions from matching the PICO elements, the strength of evidence of the citation, and matching of the clinical task that generated the question and the task orientation of the citation:

$$S_{EBM} = \lambda_p S_{PICO} + \lambda_s S_{SoE} + \lambda_t S_{task} \tag{9.5}$$

With few exceptions, the score components were derived heuristically based on recommendations for critical appraisal of medical literature, intuition and observations on a training set.

It is safe to assume that these top-level scores are generated by three different scoring systems, and resort to fusion of the scores. Fox and Shaw (1994) explored different methods for combining scores, and showed the "sum" method to be the best fusion approach. Many fusion methods have been explored since Zhang *et al.* (2001), however, adding the scores is still a viable approach to exploration of fusion (Aronson *et al.* 2005a).

Only the S_{PICO} score needs to be adjusted with respect to the question. The S_{SoE} and the S_{task} scores derived in semantic processing contribute to the overall score without further adjustments.

9.2.3 Question–Document frame matching (PICO score)

Matching of the PICO elements in a question and in a document frame is the primary responsibility of the Semantic Matcher. This process results in the assignment of the PICO score to the document. Each extracted PICO element contributes to the score proportionally to its role. To reduce the system's susceptibility to automatic mapping errors, the most widely used surface representation(s) of a concept are used in the matching process independent of the concept's presence in the UMLS. The rules for score assignment are built into individual components scoring. There are two types of rules: global (task-independent) and local (task-specific) rules. The individual components' scores are combined linearly according to Equation 9.6.

$$SPICO = S_{problem} + S_{population} + S_{intervention} + S_{outcome} \tag{9.6}$$

Problem matching and scoring

The first component in the above equation, $S_{problem}$, depends on a match between the primary problem in the question frame and the primary problem in the abstract (the highest-scoring problem identified by the problem extractor). A score of 1 is given if the problems match exactly based on their unique UMLS identifier as provided by MetaMap. Matching based on concept identifiers provides for a conceptual match disregarding surface representation of the term used in a document. Failing an exact match of concept identifiers, a partial string match is given a score of 0.5. The string match accounts for cases in which one of the frames contains a more specific term. If the primary problem in the query has no overlap with the primary problem from the abstract, a score of −1 is given. The initial intent was to remove such citations from the set. However, given the accuracy of current tools and the incompleteness of the Metathesaurus, the hard binary constraint was replaced with demoting the documents, which resulted in better performance in the exploratory experiments. Finally, if the problem extractor could not identify a problem (but the query frame does contain a problem), a score of −0.5 is given. The primary problem matching rules are global (applied universally independent of the clinical task).

Co-occurring problems must be taken into consideration in the *differential diagnosis* and *etiology* tasks because knowledge of the problems is typically incomplete in these scenarios. Therefore, physicians might be interested in any problems mentioned in the abstracts in addition to the primary problem specified in the query frame. The local rules for *differential diagnosis* and *etiology* questions give three points to disorders mentioned in the title, and disorders mentioned anywhere else receive one point (in addition to the match score based on the primary problem).

Population matching and scoring
The population score is global. It is based on the premise that a question frame can contain only one description of a patient. If the patient description matches the population identified in a document, the document score is incremented by one. For example, finding the question population group *children* in the document population slot increments the match score by one. There is no penalty for not matching the patient slot.

Intervention matching and scoring
According to the global scoring rules, for each intervention in the question frame that matches an intervention in a document intervention slot, the intervention score is incremented by one. The intervention score is then normalized (divided by the number of interventions in the question frame), and added to the document score. If no intervention in the question frame matches interventions in the document frame, a score of –0.5 is given. For therapy and diagnosis questions with empty intervention slots, a score of one is given to documents in which interventions with an appropriate semantic type were identified.

Outcome scoring
The outcome score is not based on frame matching even if the desired outcome of the intended intervention is specified in the question frame. This decision is based on the analysis of real life questions that mostly do not specify a high level outcome. More importantly, it is highly unlikely that a clinician will choose a surface representation of the desired outcomes that will match those in the article literally, and semantic matching on the outcomes level requires deeper understanding and reasoning about outcomes than is currently available.

Rather then forgo the outcome statements, the score of the highest-ranking outcome sentence generated by the outcome extractor is added to the document score. This decision is motivated by the assumption that outcome statements contain answers to questions, or at least present enough information to predict whether an answer could be found in the text of the article. Given a match on the primary problem and other elements, all highly ranked patient outcomes are likely to be of interest.

9.2.3.1 *Answer generation*

The CQA-1.0 system generates answers of two types: (1) multi-tiered answers and (2) best answers. The goal of the multi-tiered answer generated by the CQA-1.0 system is to provide an overview of available information, which is appropriate for the majority of clinicians' information needs. However, there are cases when an overview is not needed. This situation is similar to a known item search. In general, when searching for a known item, the user knows of a particular document, but does not know where it is. In clinical practice that could actually be the case, but more often clinicians want to verify that their recollection of a fact is correct. For example, they might want to verify that there is a contraindication for a generally accepted treatment in a certain group of patients. In such situations generating a short list of answers or a single best answer is appropriate.

The semantic representation of each document and the EBM-based ranking described in Section 9.2.2.7 provide a means for extracting and presenting both answer types. A TREC QA model is used to generate answers for a known fact confirmation: outcome statements from top *N* documents are extracted as a short list of likely answers to the best of the system's knowledge. In this case, answer generation amounts to displaying the title and the annotated outcome statements from the top ranking documents, since all information is already available in the CQA-1.0 document frames.

For example, the top-ranking answer for the question about constipation in the cancer patient above is found in MEDLINE citation (PMID 18454610).

An overview answer is generated using clustering – a method known to provide a good overview of data and often used to visualize and interactively explore large document collections and knowledge bases (Card, Mackinlay, & Shneiderman 1999). To generate this answer type, documents discussing the same intervention, or interventions belonging to the same drug class are identified. A list of intervention classes serves as the top-tier answer. Each intervention is supplemented with supporting evidence (the outcome statement extracted from the top ranking citation in the cluster), and with the ranked list of citations in the cluster. This approach provides a full answer that is hypothesized to be better suited for the domain.

9.2.4 Semantic clustering

The retrieved MEDLINE citations are organized into semantic clusters based on the main interventions (interventions with top scores) identified in the abstract text by the intervention extractor, and using hierarchical agglomerative clustering based on the UMLS hierarchical relationships (Demner-Fushman & Lin 2006a; Demner-Fushman & Lin 2006b). The clustering process starts by placing each concept in its own group (with N identified interventions, each in its own cluster).

Iteratively, interventions that fall under a common parent (a UMLS hypernym) are grouped together, ascending the UMLS hierarchy in the process. For example, rofecoxib would be grouped with ibuprofen because they are both Anti-Inflammatory Agents according to UMLS.

The process is applied until no new clusters can be formed. In order to preserve granularity at the level of practical clinical interest, the tops of the UMLS hierarchies were manually truncated. For example, the MeSH category *Chemical and Drugs* was eliminated as too general. The most general ancestor concept, which often represents a drug class, for example, antibiotics, is then used as the cluster label.

Summary

The question answering process implemented in the CQA-1.0 system follows these steps:

- Clinical questions are translated to question frames and search queries.
- MEDLINE documents are retrieved and translated to CQA-1.0 document frames.
  - Knowledge extractors fill the population, problem(s), intervention(s), and outcome slots of the document frame.
  - Each clinical scenario element is scored, and preserves a pointer to its position in the citation.
  - The result of the semantic processing is a set of document frames fully annotated with the elements of the three components of the semantic domain model: PICO, Clinical Task, and Strength of Evidence.
- The semantic matcher module (see Section 9.2.3) compares the document frame with the question frame. The scoring and ranking of the document frames in the semantic matcher concludes the processing of the documents retrieved to answer the question.
- The desired form of the answer determines the next processing step:
  - For the best answer, the title of the abstract and the top three outcome sentences in the order they appeared in the abstract are combined, then the answers from n-best citations are returned without any further processing.
  - Multi-tiered answers are generated using hierarchical agglomerative clustering of the annotated citations described in Section 9.2.4.

CHAPTER 10

Software engineering

10.1 Introduction

In 2006, Geoffrey Chang was a rising star of the structural biology world. Determining the structure of a protein experimentally is often very difficult or impossible, so there is a thriving research field in the computational prediction of protein structure.

Chang worked on a small but important set of molecules that help determine things like development of antibiotic resistance and individual responses to chemotherapy for cancer. This led to his work being quite influential. It was difficult to get grants funded if one's preliminary results seemed to conflict with Chang's work, and it was difficult to publish papers that seemed to conflict with Chang's findings. However, a growing body of evidence amassed that suggested that there was a problem with Chang's predictions. Chang went looking for the source of the discrepancy and discovered that there was a simple bug in one of his software programs. It caused the signs of the numbers in two columns of data to be reversed, so that positive numbers became negative and negative numbers became positive. This caused the reversal of a crucial element of a protein's structure, known as its handedness (Miller 2006).

The consequences of Chang's bug for other people have been described above. The consequences for Chang were the retraction of five papers in some of the most prestigious journals in the world, including *Science* and *Nature* (Chang *et al.* 2006). The consequences for human health and the battle against disease are unknown. The Chang saga illustrates the often-ignored fact that good software testing is not just important for commercial products – it is essential for academic systems, as well. As Robin Knight has put it, "For scientific work, bugs don't just mean unhappy users who you'll never actually meet: they mean retracted publications and ended careers. It is critical that your code be fully tested before you draw conclusions from results it produces" (personal communication). (On a related topic, it is good practice for any scientific software to be made available open source, both for reasons of public scrutiny and of reproducibility (Pedersen 2008).)

Good testing is especially important in the field of biomedical NLP. In theory, mainstream NLP is concerned with developing testable hypotheses about natural language processing or about the human language faculty. In contrast, a large percentage of the work in BioNLP is explicitly motivated by the desire to build tools for working biologists and clinicians. Nonetheless, it seems clear that a substantial proportion of BioNLP software is not subjected to industrial-strength testing. Indeed, one survey of web sites offering BioNLP services or data extracted by text mining showed evidence of not having been subjected to the most simple software test imaginable (Cohen, Hunter, and Palmer 2013).

In this chapter, we will explore three related but distinct topics. The first is software testing in general; the material in this section of the chapter could be applied to any sort of software program. The second is usability testing, focussing on search interfaces. The third is specific to natural language processing applications, and has to do with white-box testing of text mining applications, or the granular evaluation of the ability of systems to handle particular sorts of linguistic phenomena.

10.2 Principles

Software testing should be approached with some general principles in mind. These include the following:

1. There is no such thing as bug-free software.
2. For any non-trivial software application, there is an infinite number of tests that can be run. Part of the art and science of software testing is to figure out the finite set of tests that can be run with the time and money resources available that has the highest likelihood of uncovering bugs.
3. Testing should include unexpected inputs.
4. Testing requires planning.

10.3 General software testing

Software testing is sometimes discussed as having four different levels of "maturity:"

1. Software testing is not thought about at all.
2. Software testing is demonstrating that the software works.
3. Software testing is demonstrating that the software doesn't work.
4. Software testing is fully integrated into the development process, driving all aspects of software construction.

We will define software testing somewhere in the region of maturity levels 3 and 4. For our purposes, software testing is the science and art of finding bugs in computer software and supporting data. We will define a *bug* as any deviation between the actual performance and the intended performance of a software program. Bugs may be minor, such as a mis-spelling in a user interface or error message, or critical, such as the molecular handedness bug discussed above. However, all types of bugs should be searched for.

It should be noted that there is no such thing as a non-trivial program that contains no bugs. No software should ever be considered to be 100% free of bugs, whether it is an expensive enterprise commercial software product or a piece of code written for a class project. However, this fact does not relieve us of the responsibility to search diligently for bugs in our code, nor does it preclude the possiblity of us finding the vast majority of them, given sufficient attention to the testing process.

10.3.1 Clean and dirty tests

One of the consequences of our conception of software testing is that a test suite, or a set of tests for a particular application, should include both *clean tests*, which consist of inputs that the program should be expected to receive and should be expected to handle and which can be thought of as "demonstrating that the software works," and *dirty tests*, which consist of inputs that the system is not expected to receive. It has been claimed that mature testing organizations have a ratio of 5:1 of dirty tests to clean tests, while immature testing organizations have a ratio of 5:1 of clean tests to dirty tests. This is not to imply that test suites should not contain clean inputs, but rather that dirty inputs cannot be ignored.

10.3.2 Testing requires planning

Testing should be planned, just as development is. As a good rule of thumb, it is a safe bet that if you do not plan to test something, you will not test it. Test-planning is a useful exercise for a variety of reasons. The most obvious reason is that it helps to guide the testing effort. Another is that it may help to uncover gaps in the definition of requirements for the program. (For example, what is the desired behavior if a call to an information retrieval module returns no documents?) A very important one is that it may help identify system components that have been designed in a way that makes them difficult or impossible to test. Such components are highly likely to contain bugs and are prime candidates for redesign.

Although test plans in industry may be book-length, a test plan does not have to be long or elaborate to be useful, and even a two-page test plan may have salutary effects on a project. Short test plans may even have an advantage in that stakeholders are more likely to read them thoroughly. In general, test plans should be reviewed by all stakeholders in a project, including the project manager, developers, and other testers. In an industrial context, review of a test plan by sales and marketing personnel may reveal important disconnects between the engineering team and the sales and marketing teams with respect to the expected scope of a program. In an academic context, review of a test plan may reveal important differences between the advisor's vision for a program and the student's.

Just as important as covering what will be tested can be specific mention of what will not be tested. (One of the authors wrote a test plan in an industrial context that specified that the security functions of the database component of an application could not and would not be tested with available resources. The marketing department insisted on having the database functionality developed nonetheless. Soon after release the product was found to have security bugs in the untested database component, causing withdrawal of the functionality from a popular website at the cost of much embarrassment and a great amount of wasted development time.)

10.3.3 Catalogues

One of the first steps in building a test plan is typically to list the types of inputs that a program might receive. This can be more difficult than it might seem at first glance. To get a flavor for the nature of the problem, try this exercise[13]. Taking a non-linguistic example for simplicity's sake, consider an application of the following sort. It takes three input integer values representing the lengths of the sides of a triangle and determines whether the triangle is scalene, isosceles, or equilateral[14]. Pause now and list all of the inputs that you can think of for such a program. Then flip forward a few pages, where you will find a list of such inputs. Compare your list to the list in this book – how many occurred to you?

10.3.4 How many tests are possible?

We made the claim earlier that any non-trivial application can be subjected to an infinite number of tests. For an obvious example of how this can be true, consider that any application with a loop that executes under conditions like while (true) might, in theory, be capable of executing an infinite number of times. And, sometimes high numbers of repetitions of a test may be necessary to uncover a bug.

13. This famous exercise is due to Glenford J. Myers's classic work *The art of software testing*.

14. Informally: A scalene triangle has three sides of unequal length and three unequal angles. An isosceles triangle has two sides of equal length and two angles of equal length. An equilateral triangle has three sides of equal length and three equal angles of 60 degrees.

Consider, for example, a program with a memory leak; the leak might not become obvious until after many repetitions through a loop. Consider also an application that maintains the state of some variable containing an integer from call to call. The program may malfunction when a certain value for the variable is exceeded, but it might take an enormous amount of repetitions to reach that value.

The claim that an infinite number of tests is possible may seem extreme, but it is not difficult at all to come up with cases of applications that could be subjected to a mind-boggling number of tests. Consider the popular machine learning application SVMLight (Joachims 1999). SVMLight offers a large number of command-line options (see Figure 10.1 for just a partial list). A common type of software bug is that some option *A* will not work correctly just in case some other option *B* is set to a certain value. Suppose, then, that the author of SVMLight wanted to assure himself that all possible command-line option settings work with all other command-line option settings. Suppose furthermore that we make some simplifying assumptions. Assume, for example, that any option that can take unrestricted numeric inputs will only be tested with up to three values – one positive, one zero, and one negative. Even with a number of such simplifying assumptions about commands that can take unrestricted textual or numeric values, it would take 92,980,917,360 tests to cover all possible sets of combinations. If we knew what the correct behavior should be for each set of combinations and if it took only one second to run each test, this would require 2,948 years. This is not an extreme example. Again, the art and science of software testing is to figure out how to reduce this to a finite set of tests that has the highest probability of actually uncovering bugs. (For an approach to making this kind of problem tractable, see the section on combination testing using the all-pairs technique in Kaner, Bach, & Pettichord 2002.)

10.3.5 Equivalence classes

Equivalence classes are sets of inputs that all are likely to uncover the same bug. Examples of equivalence classes for numbers might include positive numbers, negative numbers, zero, integers, and reals. Note that any one input might belong to more than one equivalence class. For example, +0.0 is a signed rational zero. Determining the relevant set of equivalence classes for an application is important both because it helps us to define the types of test cases that we will need to include and because it helps us to lower the number of tests that we must run. For example, if we have a function that takes a numerical value as input, then we know that we must run tests that cover each equivalence class. However, we do not need to run every number in that equivalence class – once we have determined the

```
Learning options:
        -z {c,r,p}   - select between classification (c), regression (r), and
                       preference ranking (p) (see [Joachims, 2002c])
                       (default classification)
        -c float     - C: trade-off between training error
                       and margin (default [avg. x*x]^-1)
        -w [0..]     - epsilon width of tube for regression
                       (default 0.1)
        -j float     - Cost: cost-factor, by which training errors on
                       positive examples outweight errors on negative
                       examples (default 1) (see [Morik et al., 1999])
        -b [0,1]     - use biased hyperplane (i.e. x*w+b0) instead
                       of unbiased hyperplane (i.e. x*w0) (default 1)
        -i [0,1]     - remove inconsistent training examples
                       and retrain (default 0)
Performance estimation options:
        -x [0,1]     - compute leave-one-out estimates (default 0)
                       (see [5])
        -o ]0..2]    - value of rho for XiAlpha-estimator and for pruning
                       leave-one-out computation (default 1.0)
                       (see [Joachims, 2002a])
        -k [0..100]  - search depth for extended XiAlpha-estimator
                       (default 0)
Transduction options (see [Joachims, 1999c], [Joachims, 2002a]):
        -p [0..1]    - fraction of unlabeled examples to be classified
                       into the positive class (default is the ratio of
                       positive and negative examples in the training data)
Kernel options:
        -t int       - type of kernel function:
                        0: linear (default)
                        1: polynomial (s a*b+c)^d
                        2: radial basis function exp(-gamma ||a-b||^2)
                        3: sigmoid tanh(s a*b + c)
                        4: user defined kernel from kernel.h
        -d int       - parameter d in polynomial kernel
        -g float     - parameter gamma in rbf kernel
        -s float     - parameter s in sigmoid/poly kernel
        -r float     - parameter c in sigmoid/poly kernel
        -u string    - parameter of user defined kernel
Optimization options (see [Joachims, 1999a], [Joachims, 2002a]):
        -q [2..]     - maximum size of QP-subproblems (default 10)
        -n [2..q]    - number of new variables entering the working set
                       in each iteration (default n = q). Set n<q to prevent
                       zig-zagging.
        -m [5..]     - size of cache for kernel evaluations in MB (default 40)
```

Figure 10.1 A partial list of the command-line options for SVMLight.

equivalence class, we will typically pick a small number of representatives from that class. Thinking about intersections between equivalence classes is helpful. For example, if we know that real numbers are an equivalence class and that negative numbers are an equivalence class, we will probably want to make sure that our test cases include some real numbers that are negative numbers and some that are not. Partitioning our data in this way helps us to interpret it a priori.

10.3.6 Boundary conditions

Boundary conditions or edge effects concern the limits of the ranges at which our program is expected to operate. For example, a program may be designed to go through a loop anywhere from one to a thousand times; the boundaries are then one and one thousand. Bugs tend to "congregate" around boundaries, for a variety of reasons. For one thing, requirements may be unclear in boundary cases. For another, off-by-one errors are most evident at boundaries. Testing boundary conditions should include not just the boundaries themselves, but values that cluster around them. For example, for the hypothetical program with the loop that is expected to execute from one to one thousand times, we should try to make it execute:

- zero times
- once
- twice
- 999 times
- 1,000 times
- 1,001 times

Boundary condition errors are extremely commonplace. One is reported in Gao & Vogel (2008). While building a parallelizable version of the GIZA++ word alignment package, they first had to fix a bug that they described like this:

> When training the HMM model, the matrix for the HMM trellis will not be initialized if the target sentence has only one word. Therefore some random numbers are added to the counts. This bug will also crash the system when linking against [the] *pthread* library. We observe different alignment and slightly lower perplexity after fixing the bug.

One of the authors has made a boundary condition bug so often that he names it after himself when teaching introductory programming classes. "The Cohen bug" consists of relying on item separators to recognize the end of an item when reading in items from a data file. If the last item is not followed by an item separator, as it usually isn't, then the last item in the file is not read in – a boundary error or edge effect.

10.4 Code coverage

How do we know when we have done enough testing? This is related to the question of how we know how much testing we have done. One answer to this question can certainly be derived from our test plan. One popular maxim suggests that you "test until you don't feel nervous any more." One popular approach is to measure code coverage. The term *code coverage* refers to the amount of code that is executed when a test suite is run. There are various things that can be measured, with varying levels of granularity. Line coverage is the percentage of lines of code in a program that are executed. A more granular measure is the percentage of branches of conditionals that have been followed. A less granular measure is the number of methods or classes that have been tested.

One might think that the best way to ensure high code coverage is to process large amounts of data. Intuitively, it might seem to be the case that processing a large corpus would be the most effective way of achieving high code coverage, since a large corpus is likely to contain most types of inputs. However, Cohen, Baumgartner Jr., & Hunter (2008) showed that this is not the case. They compared the code coverage achieved by running an application against the largest biomedical corpus currently available and by running it against a small, planned test suite. As Table 10.1 shows, they found that by every measure of code coverage, coverage was higher when using the test suite than when using the very large corpus. Furthermore, as soon as they started to try to increase code coverage, they uncovered two showstopper bugs.

Table 10.1 Application and package-level coverage statistics using the developer's structured tests and the full corpus. Data is given for the full application and for the two main packages in the program, one of which parses inputs and the other of which represents semantic rules. The highest number in each row is bolded. Note that coverage is always higher with the structured tests.

Metric	Functional tests	Large corpus
Overall line coverage	**56%**	41%
Overall branch coverage	**41%**	28%
Parser line coverage	**55%**	41%
Parser branch coverage	**57%**	29%
Rules line coverage	**63%**	42%
Rules branch coverage	**71%**	24%

Testing without monitoring code coverage typically results in only 50–60% of the code being covered (McConnell 2004, p. 526, citing Wiegers 2002). It is instructive to note that even with the structured test suite, only 56% total line coverage was achieved (see table), as predicted, but with the corpus, only 41% was achieved – much lower than typical coverage for a structured testing effort.

10.5 When your input is language

Building a catalogue of test conditions is relatively easy for some types of data. For example, Table 10.2 gives a catalogue of test conditions for when the input is numeric, for a program that is supposed to accept any integer from 1 to 99, adapted from (Kaner, Nguyen, & Falk 1999). However, it is often less obvious how to create a catalogue of test conditions for when your input is linguistic in nature. (We leave aside here the issues of character encodings and the like.) In this case, a helpful approach is to think of the computer program like a linguist, and in particular, to think of it as an unknown language that you are trying to describe. Known as field or descriptive linguistics, this involves determining the relevant elements of the input(s) and how they can combine. To do this, it will often be helpful to seek the input of an actual field linguist.

Table 10.2 A catalogue for a program that accepts any integer between 1 and 99. Adapted from Kaner, Nguyen, & Falk (1999).

"clean" tests	"dirty" tests
any number between 1 and 99	any number less than 1
1 (boundary condition)	0 (boundary condition)
99 (boundary condition)	negative integers
	any number greater than 99
	100 (boundary condition)
	any rational number
	any non-number

Determining boundary conditions and equivalence classes and the set of test catalogues in the case where the input is linguistic may not be obvious. One prevalent boundary class is length. For example, one gene mention system that we tested did a good job at finding gene mentions except in the case where they were a name (as opposed to a symbol) and were a single word long, e.g. *insulin, myoglobin, urease, hemoglobin,* etc. One system for recognizing mentions of concepts from an ontology in text worked well for chemical names except for single letters, which were often mistaken for amino acids. Above we described a bug in the GIZA++ word alignment package which occurred only with single-word sentences.

In determining equivalence classes, it may be helpful to write out a list of all of the characteristics that the linguistic unit of interest can have. For example, in the case of a part of speech tagger, this would begin with a listing of all of the parts of speech. Then one might create a list of all of the orthographic/typographic shapes that a word can have, such as:

- Combinations of character case
- Presence or absence of numerals in a word
- Presence or absence of punctuation marks in a word (e.g. *cross-linked, won't*; this will interact with the tokenization routine, of course)

Another list might include faux morphological variants, such as:

- Non-plural words that end with *s*
- Non-present participles that end with *ing*
- Non-past tense words that end with *ed*

All of these will interact with the question of whether the words in question are in-vocabulary or out-of-vocabulary for the tagger's model, if it has one.

The general topic of test suite construction for biomedical named entity recognition systems of various types has been examined in Cohen *et al.* (2004) and Cohen *et al.* (2010c). One of the questions raised in the latter is whether there are general principles of linguistic test suite construction, or whether all test suite construction efforts must be approached de novo. The experiences of field linguists suggest that at least within a family of applications, there should be common principles applicable to building any test suite. Cohen *et al.* (2010) looked in particular at named entity recognition systems for gene names and for concepts from the Gene Ontology. They found that there were a small core of feature types or equivalence classes that were common to the problem of testing applications for both types of named entities, at least in the biomedical domain. These were:

- Length
- Numerals
- Punctuation
- Function/stopwords
- Canonical form in source
- Syntactic context
- Source or authority

Note that with the exception of the last item, all of these features, which each define equivalence classes, are linguistically motivated. This underlines the importance of linguistic analysis in test suite construction for natural language processing.

From a machine learning context, a natural question to ask is whether or not analysis of your test data in this way will affect your feature engineering. The answer is probably yes, and that this is a good thing. Two desirable effects from this perspective are that:

- the test suite can then be used to help test your feature extractors independent of the learning algorithm, and
- the exercise might lead you to think of features that you wouldn't have thought about otherwise.

The next question would be whether the features that you are testing occur in naturally occurring text in sufficient quantity to make it worth reserving probability mass for them. In the authors' view, this is not necessarily a relevant question in a biomedical context. If one's only goal is to maximize performance on some shared task, then certainly one does not want to disproportionately allocate probability mass for rare events. However, in real biomedical applications, we are often strongly concerned with rare events, and want to be sure that we do not miss them. To give a clinical example, most people presenting in the emergency room with abdominal pain do not have appendicitis, but we never want to miss a single true positive case.

10.6 User interface evaluation

Usability may be the most important factor affecting adoption of BioNLP applications by their intended users, making the difference between a development effort that results in a widely used tool or one that languishes in disuse despite good performance on NLP tasks. Usability in BioNLP has been studied mostly in the area of search interfaces.

One of the basic principles of user interface design is that interfaces should be designed with the active involvement of potential representative users (Hearst 2009). Techniques for search usability testing range from very formal approaches on the one hand to what has been termed "discount usability testing" (Nielsen 1989) on the other. Informal usability tests can be carried out with relatively low effort compared to formal tests. Examples of approaches to informal testing include preparing paper prototypes of an interface and asking users to simulate their usage. A succession of "screens" can be prepared to simulate the effects of user actions (Hearst 2009). Another form of "discount usability testing" is known as *heuristic evaluation*. In this technique, a design expert reviews the design using a checklist of user interface desiderata. It is useful to employ multiple testing techniques, since different techniques have been shown to uncover different kinds of problems (Hearst 2009).

Hearst (2009) gives a number of principles of usability testing for search interfaces that are probably applicable to language processing tool user interfaces in general. These include:

- **Avoid experimenter bias.** Experimenter bias can be communicated to users by such mistakes as informing the user that they are testing a system that the experimenters built – this may bias the users towards wanting to please the experimenters by rating the system highly. Instead, Hearst recommends telling the subject that a number of systems are being evaluated.
- **Avoid using the same query twice with the same user.** Participants learn about topics as they are repeatedly exposed to them, and this learning effect can bias the outcome of the study.
- **Measure participant preferences.** What users do and don't like should be measured, since "interfaces that are not liked will not be used, given a choice" (Hearst 2009, p. 44). Participant preferences are typically measured using a Likert scale, in which the user gives a value on a numerical scale where values range from one end of some scale to another, e.g. "dislike strongly" to "like a lot."

Hearst and colleagues have carried out a number of studies on search user interfaces for the biomedical domain. In Hearst *et al.* (2007b), they investigated the hypothesis that allowing users to search figure captions explicitly would be helpful. They built a search engine that searched captions and then showed the corresponding figure when there was a match. Study participants were asked to rate two varieties of the interface using a Likert scale that asked whether they thought the search application was useful, non-cluttered, easy to browse, interesting, enjoyable, and non-overwhelming. On a scale from 1 to 7, where one was "strongly disagree" and 7 indicated "agree," seven out of eight participants indicated that they thought that the application was useful with a score of 6 or higher. Participants also filled out a standardized questionaire that addressed issues such as whether or not they would like to use either variety of the search engine in their work, and how they thought it compared to search via PubMed.

10.6.1 API interface usability

An understudied topic is the issue of application programming interface usability. It is likely, although it has not been experimentally demonstrated, that the ease of use of an API will affect the adoption of a tool. An example of this is the ABNER named entity recognition system. It has models that allow it to tag either gene names, or the full set of semantic classes from the JNLPBA competition. ABNER does not score nearly as well as many published systems on the BioCreative and JNLPBA metrics, and yet it is heavily used by others and popular. We hypothesize that this is because it is well-engineered, being distributed as a Java .jar file with a very simple API.

10.6.2 *Answers to the exercise*

(Myers 1979, p. 2)

1. A valid scalene triangle. (Note that this would not include all possible sets of three non-zero integers, since there are no triangles with sides of e.g. 1, 2, and 3.)
2. A valid equilateral triangle.
3. A valid isosceles triangle.
4. At least three test cases defining isosceles triangles with all possible permutations of the two equal sides.
5. A length of zero for at least one side.
6. A negative integer for at least one side.
7. Three integers greater than zero such that the sum of two of the lengths is equal to the third. (If this input returned a classification of isosceles, that would constitute a bug.)
8. At least three test cases of the preceding type covering all permutations of the location of the side equal to the sum of the other two sides.
9. Three integers greater than zero such that the sum of two of the sides is less than the length of the third.
10. All three permutations of the preceding test case.
11. All three sides are zero.
12. A test case with at least one non-integer value.
13. An input with the wrong number of input values.
14. Is there a specified output for every one of the test cases listed above?

Highly experienced professional software programmers come up with about 7.8 of these fourteen tests, on average (Myers 1979, p. 3). The typical person does much more poorly. Consider how many test cases are necessary to test this trivial program, and imagine how many it would take to test the 83,634 lines of code in the popular MetaMap software application.

CHAPTER 11

Corpus construction and annotation

11.1 Corpora in the two domains as driving forces of research

One large difference between the biological domain and the clinical domain has been the availability of corpora in the biological domain and the lack of availability of corpora in the clinical domain. This difference in availability has almost certainly been one of the critical differences, and perhaps the most important difference, leading to progress and general level of activity of research in the two disparate areas: research in the genomic BioNLP area has been made possible by a large and growing number of corpora and document collections, while research in the clinical domain has been hampered by the lack of availability of corpora and document collections.

This difference is due to legal and ethical issues that affect data in the two domains quite differently. Data in the biological domain comes exclusively from published sources, which by their nature are completely open to inspection; in addition, PubMed/MEDLINE has made enormous amounts of such material freely available. In contrast, clinical data is private, and that privacy is enshrined both by law (see the material in Chapter 2 on HIPAA) and by ethical considerations. (Note that there is a difference here between *publications* on clinical data, which are freely publicly available and were the subject of an early publicly released document collection called OHSUMED, and data from patient records, which are private and are protected by legal and ethical considerations. When we speak of clinical data in this chapter, we are referring to patient records.)

This difference in the availability of corpora has had a visible effect on the progress of research in the two domains of BioNLP. Fueled by the availability of data, research in genomic BioNLP has experienced rapid and continuing growth since its inception in 1998. In contrast, research in clinical NLP has proceeded much more slowly from its inception to the present day.

11.2 Who should build biomedical corpora?

Some of the most influential corpora in NLP have been built from newswire text. The Penn Treebank (Marcus, Marcinkiewicz, & Santorini 1993) and PropBank (Palmer, Kingsbury, & Gildea 2005), for example, were built on data from the Wall Street Journal newspaper. Earlier influential corpora such as the Brown corpus

(Kucera, Francis, & Carroll 1967) were built from a variety of mostly general English genres, and corpora that have been built for corpus linguistics are, again, mostly built from general English materials. All of these have in common the fact that no special domain expertise (other than linguistic) is required for corpus annotation. The BioNLP domain stands in strong contrast to this, in that two types of expertise are required to build corpora for these domains: linguistic expertise on the one hand, and biomedical expertise on the other. Pustejovsky and Stubbs (2013) refer to this as a multi-model annotation task. It is unusual to find both of these in a single person, let alone to be able to build a team of people with expertise in both areas. However, successful corpora have been built by combining personnel from both areas to do the annotation tasks. A successful recipe seems to be to allocate tasks to the appropriate experts, while facilitating communication between both groups. For example, the GENIA corpus was built by both biologists and linguists. The CRAFT (Colorado Richly Annotated Full Text) corpus was built by using linguists for treebanking, biologists for marking up a wide variety of named entities, and both linguists and biologists for annotating coreference resolution (Cohen *et al.* 2010b; Verspoor *et al.* 2012; Bada *et al.* 2012). Frequent formal and informal communication allowed for resolution of questions by either group. For example, biologists doing coreference resolution annotation often had questions about what counted as appositives, and linguists doing treebanking often had questions about things that occurred in parentheses, such as names of mutant strains, numbers referring to positions in a sequence, and sequences themselves.

11.3 The relationship between annotation of entities and annotation of linguistic structure

One early observation of most biomedical corpus builders is that it is important to mark up named entities first, and then eliminate them from further annotation of certain kinds, particularly syntactic structure. Consider the by-now-familiar example of *breast cancer associated 1*. The syntactic structure of this name is a noun phrase containing an adjectival phrase followed by a numeral. Although there is evidence that such "static entities" can be useful targets for information extraction (Pyysalo *et al.* 2009), it is not clear that there is value in repeatedly parsing difficult gene names. For this reason, a common tactic is to mark up named entities first and then exempt them from further analysis.

This has a reflection in language processing, as well. We have pointed out elsewhere the paradox that language processing tasks are facilitated by marking up named entities first, but that finding named entities is facilitated by language processing tasks including part of speech tagging, shallow syntactic analysis, etc.

11.4 Commonly used biomedical corpora

11.4.1 GENIA

The GENIA corpus consists of a collection of 2,000 abstracts about human blood cell development annotated with structural and linguistic information. It has probably been the single most influential resource in genomic NLP.

A striking difference between the GENIA corpus and most biomedical corpora is that GENIA has been marked up with not just semantic/biological information, but with aspects of linguistic structure. This includes tokenization, sentence boundaries, part of speech, and syntactic treebanking. All of these were manually annotated. GENIA was unique in this respect for quite a number of years.

In addition to this linguistic mark-up, GENIA was annotated with a set of genomically relevant named entities, using an ad hoc ontology built by the GENIA project. The ontology is structured as follows (taken directly from Kim *et al.* 2003). The upper levels of the ontology are:

source

- natural
  - organism
    - ⋆ multi_cell
    - ⋆ mono_cell
    - ⋆ virus
  - body_part
  - tissue
  - cell_type
  - cell_component
- artificial
  - cell_line
  - other_artificial_source

substance

- compound
  - organic
    - ⋆ amino_acid
    - ⋆ nucleic_acid
    - ⋆ lipid
    - ⋆ carbohydrate
    - ⋆ other_organic_compound
  - inorganic
- atom

other These expand further, as follows:

- amino_acid
  - protein
    - family_or_group
    - complex
    - molecule
    - subunit
    - substructure
    - domain_or_region
    - other
  - peptide
  - amino_acid_monomer
- nucleic acid
  - DNA
    - family_or_group
    - molecule
    - substructure
    - domain_or_region
    - other
  - RNA
    - family_or_group
    - molecule
    - substructure
    - domain_or_region
    - other
  - polynucleotide
  - nucleotide

The GENIA ontology, unlike the GENIA corpus, was never widely adopted outside of the GENIA project, and indeed, in the JNLPBA shared task, it was flattened considerably. It is instructive to compare it with relevant sections of the Sequence Ontology.

11.4.2 CRAFT

The Colorado Richly Annotated Full Text corpus represented the first attempt to do extensive linguistic and semantic annotation of a large collection of full-text journal articles (Cohen *et al.* 2010b; Verspoor *et al.* 2012; Bada *et al.* 2012; Cohen *et al.* 2014b). It is the only biomedical corpus that we are aware of to be specifically funded by an NIH G08 grant, rather than being a by-product of some other project. Its contents are 97 full-text journal articles from the Open Access subset of PubMedCentral. To ensure biomedical relevance, its contents were selected in coordination with the Mouse Genome Informatics project, by randomly sampling papers that had been selected by MGI for annotation. This biased the collection towards papers that discussed genes and Gene Ontology concepts.

The corpus was built over a three-year period, making it the most sustained biomedical corpus effort besides GENIA. In that time, it was manually annotated with respect to:

- Sentence boundaries
- Tokens
- Part of speech
- Syntactic treebanking
- Full coreference
- A variety of semantic classes, including Gene Ontology concepts

The coreference annotation was especially unusual. Previous efforts at doing coreference annotation in full text had mostly limited themselves to biological entities only (Gasperin 2006; Gasperin, Karamanis, & Seal 2007) (an exception was the Institute for Infocomm Research project (Yang *et al.* 2004a, b)).

CRAFT attempted to annotate full coreference of all noun phrases in the text, regardless of what they referred to. The IDENTITY and APPOSITIVE relationships were marked. The OntoNotes guidelines were used, with the exception that it was quickly concluded that there were no generics in this data in the OntoNotes sense of that term, but rather that what the OntoNotes project considered generics were actually named entities.

11.4.3 BioCreative gene mention corpora

The first and second BioCreative shared tasks resulted in the release of two gene mention corpora that came to be very widely used for the gene mention task, both as a result of their use in the shared tasks and, due to their free accessibility and the availability of baseline scores to compare against, in subsequent research on the gene mention problem.

These corpora featured tokenization, part of speech annotation (not manually corrected, unlike the case of GENIA and CRAFT), and manual annotation of all gene mentions. They consist of isolated sentences.

Unlike other gene mention corpora, the BioCreative corpora made a concerted effort to deal with the fact that there may be multiple valid boundaries for any gene mention in text. This was done by providing an auxiliary file with boundaries that differed from those in the main body of the corpus but that were considered equally correct. The annotation guidelines discuss the different notions of correctness for gene mention boundaries.

11.4.4 AIMed

The AIMed corpus (Bunescu *et al.* 2005) consists of a set of 980 abstracts, divided into three somewhat differently annotated sets.

- 750 abstracts were tagged with gene and protein names.
- 200 abstracts were tagged with gene and protein names and with over 1,000 protein interactions.
- 30 abstracts intended to serve as negative examples, containing more than one gene mention but with no discussion of gene interactions, were tagged with gene and protein names.

11.4.5 Word sense disambiguation

The National Library of Medicine's word sense disambiguation corpora are the only extant biomedical corpora for word sense disambiguation. Weeber, Mork, & Aronson (2001) describes a corpus that contains 5,000 sense-annotated UMLS terms. It is a targetted WSD corpus, focussing on 50 highly frequent ambiguous UMLS concepts, with 100 tokens of each term. It is probably the most carefully annotated corpus, with eight annotators having annotated every one of the 5,000 instances. The more recent MSH WSD test collection is larger and has broader semantic coverage, but is automatically constructed (Jimeno-Yepes & Aronson 2010).

11.4.6 Clinical corpora

Due to advances in de-identification of clinical narrative and community-wide efforts partially supported and funded by the National Institutes of Health (NIH), the dearth of clinical corpora described above is becoming less of a problem, at least for notes in American English. Several corpora described below became publicly available on request. The change in availability policies is reflected in the progression of the i2b2 (Informatics for Integrating Biology & the Bedside) NLP Shared Tasks: the first task, dedicated to de-identification of the provided discharge summaries and determination of smoking status, required destroying the dataset after participating in the challenge. This policy has now been reversed and the dataset is available on request along with the subsequently annotated data at https://www.i2b2.org/NLP/DataSets/Main.php.

11.4.6.1 *NLP Challenge*

The NLP Challenge data set (Pestian *et al.* 2007) is a classified document collection, rather than a true corpus per se. It is included here because it is the first publicly available set of clinical documents. (An earlier set of pathology reports was withdrawn for legal reasons.) It consists of a set of 1,954 radiology reports that have been labelled with ICD-9-CM codes. Adding linguistic annotation to turn this document collection into a true corpus would be a significant contribution to BioNLP.

11.4.6.2 *The MIMIC collection*

The MIMIC II (Multi-parameter Intelligent Monitoring for Intensive Care) Database is an ongoing effort of the Harvard-MIT team guided by Professor Roger Mark from the Laboratory for Computational Physiology at the Massachusetts Institute of Technology (Saeed *et al.* 2011). The database contains ICU patients' clinical information (lab results, fluid balance, medications, etc.) and, most importantly to the NLP community, nurses' progress notes, discharge summaries and radiology reports for over 22,000 patients[15].

Parts of the MIMIC II free-text document collection from Partners HealthCare and from Beth Israel Deaconess Medical Center were annotated on the document level (for example, determining if the discharged patient was a smoker) in the earlier i2b2 Shared Tasks. The most linguistically involved so far was the 2010 Fourth i2b2/VA Challenge evaluation, which combined a subset of the discharge summaries from the MIMIC II Database and a matching set of discharge summaries and progress notes from University of Pittsburgh Medical Center. The goal of the fourth evaluation was to identify complete noun phrases (NP) and adjective phrases corresponding to medical problems, treatments and tests. The phrases could include the first prepositional phrase (PP) following an annotated instance, but only when the PP indicated either an organ/body part or when the PP and NP could be rearranged to eliminate the PP. In addition to phrase extraction, the tasks included classification of medical problems (in the assertion task, the participants had to determine if the problems were present, absent, possible, associated with someone else, or conditional), and extraction of relations between concepts, for example "treatment improves medical problem".

15. http://mimic.physionet.org/UserGuide/UserGuide.pdf

The fourth i2b2/VA challenge collection consists of 349 clinical documents released as the training set and 477 documents in the test set. The documents have been fully de-identified and manually annotated for concept, assertion, and relation information. About 12,000 medical problems (43% of concepts) are annotated in the training set. The majority of the problems are asserted as present in patient. The second most frequent assertion is negation (about 20%). The training set provides about 1,700 examples of the most frequent relation – "Test reveals medical problem".

11.5 Factors that contribute to the success of biomedical corpora

Cohen *et al.* (2005a, b) examined the factors that make a biomedical corpus successful. Their definition of "success" was being used outside of the lab that created the corpus – some corpora are never used outside of the lab that created them, while others enjoy wide usage throughout the BioNLP community. As their primary data, they found a set of corpora that had been in existence long enough to have had the potential for wide-scale usage – that is, they eliminated the possibility that some of these corpora may not have been widely used simply because they had not been in existence for very long. They then counted the uses of these corpora *outside of the labs that created them* by doing extensive literature searches and by polling the corpus creators to see if they knew of additional uses of their corpora. They found a distinct difference in the number of uses of the various corpora – see Table 11.1.

Table 11.1 Uses of several corpora outside of the labs in which they were produced, as of 2005.

Name	Age	Uses
GENIA	6	21
GENETAG	1	8
Yapex	3	6
Medstract	4	3
Wisconsin	6	1
PDG	6	0

It was clear that the GENIA corpus was the most widely used, and therefore arguably the most useful, of these several efforts to build biomedical corpora. What contributed to its wide usage? A clue comes from examining the kinds of tasks that each corpus could be used for.

Table 11.2 details what kinds of tasks each corpus could be used to evaluate. GENIA was the only corpus with detailed, manually curated *linguistic and structural* mark-up, as opposed to semantic mark-up. Thus, it was the only corpus that could be used to develop and evaluate systems for basic linguistic pre-processing tasks, such as sentence segmentation, tokenization, and part of speech tagging. Combined with the fact that it was marked up with named entities, which as we have seen were the focus of much early work in biological BioNLP, it was an attractive resource for early BioNLP researchers.

GENIA also stood out with respect to the formats in which it was distributed. The corpora with the lowest usage rates were distributed in very idiosyncratic formats. In contrast, GENIA was made available both as embedded XML and in the one-token-per-line with whitespace-separated tags format that is used by many utilities.[16]

Table 11.2 Mark-ups for low-level and high-level tasks present in the various corpora examined in Cohen *et al.* (2005a, b). Among the low-level tasks, SS is sentence segmentation, T is tokenization, and POS is part of speech. Among the high-level tasks, NER is named entity recognition, IE is information extraction, A is acronym/abbreviation definition, and C is coreference.

Name	SS	T	POS	EI	IE	A	C
GENIA	•	•	•	•			
GENETAG				•			
Yapex				•			
Medstract				•		•	•
Wisconsn				•	•		
PDG				•	•		

The authors concluded that adding structural and linguistic information contributes to the usage and utility of a corpus, as does distributing it in widely accepted formats.

Since the time of the Cohen *et al.* studies on corpus utility, GENIA has continued to be developed. It has been syntactically treebanked (Tateisi *et al.* 2005) and marked up for coreference. It continues to be a valuable research resource.

16. At that time, standoff annotation had not yet become popular.

References

Afantenos, S.; Karkaletsis, V.; and Stamatopoulos, P. 2005. Summarization from medical documents: a survey. *Artificial Intelligence in Medicine* 33(2):157–177.

Agarwal, S., and Yu, H. 2009. Automatically classifying sentences in full-text biomedical articles into introduction, methods, results and discussion. *Bioinformatics* 25(23):3174–3180.

Ahlers, C. B.; Fiszman, M.; Demner-Fushman, D.; Lang, F.-M.; and Rindflesch, T. C. 2007. Extracting semantic predications from MEDLINE citations for pharmacogenomics. *Pacific Symposium on Biocomputing* 12:209–220.

AHRQ. 2002. Systems to rate the strength of scientific evidence. Technical Report No. 02-P0022, Agency for Healthcare Research and Quality.

Alex, B.; Grover, C.; Haddow, B.; Kabadjov, M.; Klein, E.; Matthews, M.; Roebuck, S.; Tobin, R.; and Wang, X. 2008. Assisted curation: Does text mining really help? In *Pac Symp Biocomput.*

Ando, R. K.; Dredze, M.; and Zhang, T. 2006. TREC 2005 genomics track experiments at IBM Watson. In *Proceedings of TREC 2005.*

Aronson, A. R., and Lang, F.-M. 2010. An overview of MetaMap: historical perspective and recent advances. *Journal of the American Medical Informatics Association (JAMIA)* 3(17):229–236.

Aronson, A. R., and Rindflesch, T. C. 1997. Query expansion using the UMLS Metathesaurus. In *Proceedings of the 1997 Annual Symposium of the American Medical Informatics Association (AMIA 1997)*, 485–489.

Aronson, A. R.; Mork, J. G.; Gay, C. W.; Humphrey, S. M.; and Rogers, W. J. 2004. The NLM indexing initiative's Medical Text Indexer. In *Proceedings of the 11th World Congress on Medical Informatics (MEDINFO 2004)*, 268–272.

Aronson, A. R.; Demner-Fushman, D.; Humphrey, S. H.; Lin, J.; Liu, H.; Ruch, P.; Ruiz, M. E.; Smith, L. H.; Tanabe, L. K.; and Wilbur, W. J. 2005. Fusion of knowledge-intensive and statistical approaches for retrieving and annotating textual genomics documents. In Voorhees, E. M., and Buckland, L. P., eds., *Proceedings of the Fourteenth Text REtrieval Conference (TREC 2005), November 2005, Gaithersburg, Maryland.* National Institute of Standards and Technology, pp. 36–45.

Aronson, A. R. 2001. Effective mapping of biomedical text to the UMLS Metathesaurus: The MetaMap program. In *Proceeding of the 2001 Annual Symposium of the American Medical Informatics Association (AMIA 2001)*, 17–21.

Bada, M., and Hunter, L. 2007. Enrichment of OBO ontologies. *Journal of Biomedical Informatics* 40:300–315.

Bada, M.; Eckert, M.; Evans, D.; Garcia, K.; Shipley, K.; Sitnikov, D.; Baumgartner Jr., W. A.; Cohen, K. B.; Verspoor, K.; Blake, J. A.; and Hunter, L. E. 2012. Concept annotation in the CRAFT corpus. *BMC Bioinformatics* 13:161.

Baeza-Yates, R., and Ribeiro-Neto, B. 1999. *Modern Information Retrieval.* Addison Wesley Longman Publishing Co. Inc.

Bathia, N.; Shah, N.; Rubin, D.; Chiang, A.; and Mussen, M. 2008. Comparing concept recognizers for ontology-based indexing: MGREP vs. MetaMap. Technical report, National Center for Biomedical Ontologies.

Baumgartner Jr., W. A.; Lu, Z.; Johnson, H. L.; Caporaso, J. G.; Paquette, J.; Lindemann, A.; White, E. K.; Medvedeva, O.; Cohen, K. B.; and Hunter, L. 2008. Concept recognition for extracting protein interaction relations from biomedical text. *Genome Biology* 9.

Bekhuis, T., and Demner-Fushman, D. 2010. Towards automating the initial screening phase of a systematic review. In *Proceedings of the 13th World Congress on Medical and Health Informatics (MEDINFO 2010).*

Biber, D.; Johansson, S.; Leech, G.; Conrad, S.; and Finegan, E. 1999. *Longman grammar of spoken and written English.* Pearson.

Blake, J. B. 1986. From Surgeon General's bookshelf to National Library of Medicine: a brief history. *Bulletin of the Medical Library Association* 74(4):318–324.

Blaschke, C., and Valencia, A. 2001. The potential use of SUISEKI as a protein interaction discovery tool. *Genome Inform* 12:123–134.

Blaschke, C.; Andrade, M. A.; Ouzounis, C.; and Valencia, A. 1999. Automatic extraction of biological information from scientific text: protein–protein interactions. In *Intelligent Systems for Molecular Biology*, 60–67.

BMJ Clinical Evidence. 2010. Available from: http://clinicalevidence.bmj.com/. Accessed 2010.

Booth, A., and O'Rourke, A. 1997. The value of structured abstracts in information retrieval from MEDLINE. *Health Libraries Review* 14(3):157–166.

Browne, A. C.; Divita, G.; Aronson, A. R.; and McCray, A. T. 2003. UMLS language and vocabulary tools. In *Proceedings of the 2003 Annual Symposium of the American Medical Informatics Association (AMIA 2003)*, 798.

Bunescu, R.; Ge, R.; Kate, R. J.; Marcotte, E. M.; Mooney, R. J.; Ramani, A. K.; and Wong, Y. W. 2005. Comparative experiments on learning information extractors for proteins and their interactions. *Artificial Intelligence in Medicine* 33(2):139–155.

Caporaso, J. G.; Baumgartner Jr., W. A.; Cohen, K. B.; Johnson, H. L.; Paquette, J.; and Hunter, L. 2005. Concept recognition and the TREC Genomics tasks. In *The Fourteenth Text REtrieval Conference (TREC 2005) Proceedings.*

Caporaso, J. G.; Baumgartner Jr., W. A.; Randolph, D. A.; Cohen, K. B.; and Hunter, L. 2007. MutationFinder: A high-performance system for extracting point mutation mentions from text. *Bioinformatics* 23:1862–1865.

Card, S. K.; Mackinlay, J. D.; and Shneiderman, B., eds. 1999. *Readings in Information Visualization: Using Vision to Think.* San Francisco, CA, USA: Morgan Kaufmann Publishers.

Chang, G.; Roth, C. R.; Reyes, C. L.; Pornillos, O.; Chen, Y.-J.; and Chen, A. P. 2006. Letters: Retraction. *Science* 314:1875.

Chapman, W. W.; Bridewell, W.; Hanbury, P.; Cooper, G. F.; and Buchanan, B. G. 2001. A simple algorithm for identifying negated findings and diseases in discharge summaries. *Journal of Biomedical Informatics* 24:301–310.

Chatr-aryamontri, A.; Ceol, A.; Palazzi, L. M.; Nardelli, G.; Schneider, M. V.; Castagnoli, L.; and Cesareni, G. 2006. MINT: the Molecular INTeration database. *Nucleic Acids Research* 35.

Chen, L., and Friedman, C. 2004. Extracting phenotypic information from the literature via natural language processing. *Stud Health Technol Inform* 107(2):758–762.

Chen, E. S.; Hripcsak, G.; Xu, H.; and Friedman, C. 2008. Automated acquisition of disease drug knowledge from biomedical and clinical documents: an initial study. *Journal of the American Medical Informatics Association: JAMIA* 15(1):87–98.

Chun, H.-W.; Tsuruoka, Y.; Kim, J.-D.; Shiba, R.; Nagata, N.; Hishiki, T.; and Tsujii, J. 2006. Automatic recognition of topic-classified relations between prostate cancer and genes using MEDLINE abstracts. *BMC Bioinformatics* 7.

Cimino, J. J.; Aguirre, A.; Johnson, S. B.; and Peng, P. 1993. Generic queries for meeting clinical information needs. *Bulletin of the Medical Library Association* 81(2):195–206.

Cohen, K. B., and Hunter, L. 2006. A critical revew of PASBio's argument structures for biomedical verbs. *BMC Bioinformatics* 7(Suppl. 3).

Cohen, K. B.; Baumgartner Jr., W. A.; and Hunter, L. 2008. Software testing and the naturally occurring data assumption in natural language processing. In *Software Engineering, Testing, and Quality Assurance for Natural Language Processing*, 23–30. Columbus, Ohio: Association for Computational Linguistics.

Cohen, K. B.; Dolbey, A.; Acquaah-Mensah, G.; and Hunter, L. 2002. Contrast and variability in gene names. In *Natural language processing in the biomedical domain*, 14–20. Association for Computational Linguistics.

Cohen, K. B.; Tanabe, L.; Kinoshita, S.; and Hunter, L. 2004. A resource for constructing customized test suites for molecular biology entity identification systems. In *HLT-NAACL 2004 Workshop: BioLINK 2004, Linking Biological Literature, Ontologies and Databases*, 1–8. Association for Computational Linguistics.

Cohen, K. B.; Fox, L.; Ogren, P.; and Hunter, L. 2005a. Empirical data on corpus design and usage in biomedical natural language processing. In *American Medical Informatics Association Symposium*, 156–160.

Cohen, K. B.; Fox, L.; Ogren, P. V.; and Hunter, L. 2005b. Corpus design for biomedical natural language processing. In *Proceedings of the ACL-ISMB workshop on linking biological literature, ontologies and databases*, 38–45. Association for Computational Linguistics.

Cohen, K. B.; Hunter, L.; and Palmer, M. 2014a. Assessment of software testing and quality assurance in natural language processing applications and a linguistically inspired approach to improving it. EternalS 2013, Springer, Lecture Notes in Computer Science.

Cohen, K. B.; Johnson, H. L.; Verspoor, K.; Roeder, C.; and Hunter, L. E. 2010. The structural and content aspects of abstracts versus bodies of full text journal articles are different. *BMC Bioinformatics* 11(492).

Cohen, K. B.; Lanfranchi, A.; Corvey, W.; Baumgartner Jr., W. A.; Roeder, C.; Ogren, P. V.; Palmer, M.; and Hunter, L. E. 2010. Annotation of all coreference in biomedical text: Guideline selection and adaptation. In *BioTxtM 2010: 2nd workshop on building and evaluating resources for biomedical text mining*, 37–41.

Cohen, K. B.; Roeder, C.; Baumgartner Jr., W. A.; Hunter, L.; and Verspoor, K. 2010. Test suite design for biomedical ontology concept recognition systems. In *Proceedings of the Language Resources and Evaluation Conference.*

Cohen, K. B.; Christiansen, T.; and Hunter, L. E. 2011. Parenthetically speaking: Classifying the contents of parentheses for text mining. In *Proceeding of the 2011 Annual Symposium of the American Medical Informatics Association (AMIA 2011)*, 267–272.

Cohen, K. B.; Verspoor, K.; Bada, M.; Palmer, M.; and Hunter, L. E. 2014. The Colorado Richly Annotated Full-Text Corpus (CRAFT). Multi-model annotation in the biomedical domain. In Ide, N. and Pustejovsky, J. *Handbook of Linguistic Annotation*. Springer.

Collier, N.; Park, H. S.; Ogata, N.; Tateishi, Y.; Nobata, C.; Ohta, T.; Sekimizu, T.; Imai, H.; Ibushi, K.; and Tsujii, J. 1999. The GENIA project: corpus-based knowledge acquisition and information extraction from genome research papers. In *Ninth Conference of the European Chapter of the Association for Computational Linguistics (EACL-99)*, 271–272.

Consortium, T. G. O. 2001. Creating the Gene Ontology resource: design and implementation. *Genome Research* 11:1425–1433.

Corbett, P.; Batchelor, C.; and Teufel, S. 2007. Annotation of chemical named entities. In *Biological, translational, and clinical language processing*, 57–64. Prague, Czech Republic: Association for Computational Linguistics.

Craven, M., and Kumlien, J. 1999. Constructing biological knowledge bases by extracting information from text sources. In *Intelligent Systems for Molecular Biology*, 77–86.

Czarnecki, J.; Nobeli, I.; Smith, A. M.; and Shepherd, A. J. 2012. A text-mining system for extracting metabolic reactions from full-text articles. *BMC Bioinformatics* 13:172.

Damianos, L.; Day, D.; Hirschman, L.; Kozierok, R.; Mardis, S.; McEntee, T.; McHenry, C.; Miller, K.; Ponte, J.; Reeder, F.; van Guilder, L.; Wellner, B.; Wilson, G.; and Wohlever, S. 2002. Real users, real data, real problems: the MiTAP system for monitoring bio events. In *Proceedings of BTR2002: unified science and technology for reducing biological threats and countering terrorism*.

Demner-Fushman, D., and Lin, J. 2006a. Answer extraction, semantic clustering, and extractive summarization for clinical question answering. In *Proceedings of the 21st International Conference on Computational Linguistics and 44th Annual Meeting of the Association for Computational Linguistics (COLING/ACL 2006)*.

Demner-Fushman, D., and Lin, J. 2006b. Situated question answering in the clinical domain: Selecting the best drug treatment for diseases. In *Proceedings of COLING/ACL 2006 Workshop on Task-Focused Summarization and Question Answering*.

Demner-Fushman, D., and Lin, J. 2007. Answering clinical questions with knowledge-based and statistical techniques. *Computational Linguistics* 33(1):63–103.

Demner-Fushman, D.; Mork, J. G.; Shooshan, S. E.; and Aronson, A. R. 2010. UMLS content views appropriate for NLP processing of the biomedical literature vs. clinical text. *Journal of biomedical informatics* 43(4):587–594.

Demner-Fushman, D.; Abhyankar, S.; Jimeno-Yepes, A.; Loane, R. F.; Rance, B.; Lang, F.-M.; Ide, N. C.; Apostolova, E.; and Aronson, A. R. 2011. A knowledge-based approach to medical records retrieval. In *TREC*.

Demner-Fushman, D.; Chapman, W. W.; and McDonald, C. J. 2009. What can natural language processing do for clinical decision support? *Journal of Biomedical Informatics* 42(5):760–772.

Denny, J. C.; Smithers, J. D.; Spickard, A.; and Miller, R. A. 2002. A new tool to identify key biomedical concepts in text documents, with special application to curriculum content. In *Proceedings of the 1997 Annual Symposium of the American Medical Informatics Association (AMIA 1997)*, 1007.

Divoli, A.; Wooldridge, M.; and Hearst, M. 2010. Full text and figure display improves bioscience literature search. *PLoS ONE* 5(4).

Donaldson, I.; Martin, J.; de Bruijn, B.; Wolting, C.; Lay, V.; Tuekam, B.; Zhang, S.; Baskin, B.; Bader, G.; Michalickova, K.; Pawson, T.; and Hogue, C. 2003. PreBIND and Textomy–mining the biomedical literature for protein–protein interactions using a support vector machine. *BMC Bioinformatics* 4(11).

Dowell, K.; McAndrews-Hill, M.; Hill, D.; Drabkin, H.; and Blake, J. 2009. Integrating text mining into the MGI biocuration workflow. *DATABASE: The Journal of Biological Databases and Curation*.

Du, X.-J.; Bathgate, R. A.; Samuel, C. S.; Dart, A. M.; and Summers, R. J. 2010. Cardiovascular effects of relaxin: from basic science to clinical therapy. *Nat Rev Cardiol* 7(1):48–58.

Ebell, M. H.; Siwek, J.; Weiss, B. D.; Woolf, S. H.; Susman, J.; Ewigman, B.; and Bowman, M. 2004. Strength of Recommendation Taxonomy (SORT): A patient-centered approach to grading evidence in the medical literature. *The Journal of the American Board of Family Practice* 17(1):59–67.

Elhadad, N.; Kan, M.-Y.; Klavans, J. L.; and McKeown, K. R. 2005. Customization in a unified framework for summarizing medical literature. *Artificial Intelligence in Medicine* 33(2):179–198.

Elhadad, N. 2006. *User-sensitive text summarization: Application to the medical domain.* Ph.D. Dissertation, Columbia University.

Ely, J. W.; Osheroff, J. A.; Gorman, P. N.; Ebell, M. H.; Chambliss, M. L.; Pifer, E. A.; and Stavri, P. Z. 2000. A taxonomy of generic clinical questions: classification study. *BMJ* 321:429–432.

Ely, J. W.; Osheroff, J. A.; Chambliss, M. L.; Ebell, M. H.; and Rosenbaum, M. E. 2005. Answering physicians' clinical questions: Obstacles and potential solutions. *Journal of the American Medical Informatics Association* 12(2):217–224.

Exchange, P. 2010. Parkhurst exchange. Available from: http://www.parkhurstexchange.com/searchQA. Canadian monthly GP/FP journal, accessed 2010.

Fang, H.; Murphy, K.; Jin, Y.; Kim, J.; and White, P. 2006. Human gene name normalization using text matching with automatically extracted synonym dictionaries. In *Linking natural language processing and biology: towards deeper biological literature analysis*, 41–48. Association for Computational Linguistics.

Flaherty, R. J. 2004. A simple method for evaluating the clinical literature. *Family Practice Management* 11(5):47–52.

Florance, V. 1992. Medical knowledge for clinical problem solving: a structural analysis of clinical questions. *Bulletin of the Medical Library Association* 80(2):140–149.

Fox, E. A., and Shaw, J. A. 1994. Combination of multiple searches. In *Proceedings of the 2nd Text REtrieval Conference (TREC-2)*, 243–252.

Friedman, C.; Sager, N.; Chi, E. C.; Marsh, E.; Christenson, C.; and Lyman, M. S. 1983. Computer structuring of free-text patient data. In *Proceedings of the Annual Symposium on Computer Application in Medical Care*, 688–691.

Friedman, C.; Alderson, P. O.; Austin, J. H.; Cimino, J. J.; and Johnson, S. B. 1994. A general natural-language text processor for clinical radiology. *Jornal of the American Medical Informatics Association* 1(2):161–174.

Friedman, C.; Liu, H.; Shagina, L.; Johnson, S.; and Hripcsak, G. 2001. Evaluating the UMLS as a source of lexical knowledge for medical language processing. In *Proc. AMIA Annual Symposium*, 189–193.

Friedman, C. 2005. *Semantic text parsing for patient records.* New York: Springer. Chapter 15, 423–448. Hsinchun Chen and Sherrilynne S. Fuller and Carol Friedman and William Hersh.

Fukuda, K.; Tamura, A.; Tsunoda, T.; and Takagi, T. 1998. Toward information extraction: identifying protein names from biological papers. In *Pac Symp Biocomput*, 707–718.

Gabow, A.; Leach, S. M.; Baumgartner Jr., W. A.; Hunter, L. E.; and Goldberg, D. S. 2008. Improving protein function prediction methods with integrated literature data. *BMC Bioinformatics* 9(198).

Gaizauskas, R.; Herring, P.; Oakes, M.; Beaulieu, M.; Willett, P.; Fowkes, H.; and Jonsson, A. 2001. Intelligent access to text: integrating information extraction technology into text browsers. In *Proceedings of the human language technology conference (HLT 2001)*, 189–193.

Gao, Q., and Vogel, S. 2008. Parallel implementations of word alignment tool. In *Software Engineering, Testing, and Quality Assurance for Natural Language Processing*, 49–57. Columbus, Ohio: Association for Computational Linguistics.

Garten, Y., and Altman, R. B. 2009. Pharmspresso: a text mining tool for extraction of pharmacogenomic concepts and relationships from full text. *BMC Bioinformatics* Suppl 2(10):S6.

Gasperin, C.; Karamanis, N.; and Seal, R. 2007. Annotation of anaphoric relations in biomedical full-text articles using a domain-relevant scheme. In *Proceedings of DAARC 2007.*

Gasperin, C. 2006. Semi-supervised anaphora resolution in biomedical texts. In *Linking natural language processing and biology: towards deeper biological literature analysis*, 96–103. Association for Computational Linguistics.

Guyatt, G. H.; Sackett, D.; and Cook, D. J. 1994. Users' guides to the medical literature. ii. how to use an article about therapy or prevention. b. what were the results and will they help me in caring for my patients? evidence-based medicine working group. *The Journal of the American Medical Association* 271(1):59–63.

Hafner, C.; Baclawski, K.; Futrelle, R.; Fridman, N.; and Sampath, S. 1994. Creating a knowledge base of biological research papers. In *2nd International Conference on Intelligent Systems for Molecular Biology*, 147–155.

Hakenberg, J.; Plake, C.; Leaman, R.; Schroeder, M.; and Gonzalez, G. 2008. Inter-species normalization of gene mentions with GNAT. *Bioinformatics* 24(216):126–132.

Hakenberg, J.; Gerner, M.; Haeussler, M.; Solt, I.; Plake, C.; Schroeder, M.; Gonzalez, G.; Nenadic, G.; and Bergman, C. M. 2011. The GNAT library for local and remote gene mention normalization. *Bioinformatics* 27(19):2769–2771.

Hanisch, D.; Fundel, K.; Mevissen, H.-T.; Zimmer, R.; and Fluck, J. 2005. ProMiner: rule-based protein and gene entity recognition. *BMC Bioinformatics* 6 (Suppl. 1).

Hatzivassiloglou, V.; Duboué, P. A.; and Rzhetsky, A. 2001. Disambiguating proteins, genes, and RNA in text: a machine learning approach. *Bioinformatics* 17:S97–S106.

Haynes, R. B.; Wilczynski, N.; McKibbon, K. A.; Walker, C. J.; and Sinclair, J. C. 1994. Developing optimal search strategies for detecting clinically sound studies in MEDLINE. *Journal of the American Medical Informatics Association* 1(6):447–458.

Hearst, M.; Divoli, A.; Buturu, H.; Ksikes, A.; Nakov, P.; and Wooldridge, M. 2007. BioText search engine: beyond abstract search. *Bioinformatics* 23(16):2196–2197.

Hearst, M.; Divoli, A.; Jerry, Y.; and Wooldridge, M. 2007. Exploring the efficacy of caption search for bioscience journal search interfaces. In *Biological, translational, and clinical language processing*, 73–80. Prague, Czech Republic: Association for Computational Linguistics.

Hearst, M. A. 1992. Automatic acquisition of hyponyms from large text corpora. In *Proceedings of the 14th Conference on Computational Linguistics – Volume 2*, 539–545. Morristown, NJ, USA: Association for Computational Linguistics.

Hearst, M. A. 2009. *Search user interfaces.* Cambridge University Press.

Hersh, W. R., and Greenes, R. A. 1990. Saphire – an information retrieval system featuring concept matching, automatic indexing, probabilistic retrieval, and hierarchical relationships. *Computers and biomedical research, an international journal* 23(5):410–425.

Hersh, W. R., and Voorhees, E. M. 2009. TREC genomics special issue overview. *Information Retrieval* 12(1):1–15.

Hersh, W. R.; Hickam, D. H.; Haynes, R. B.; and McKibbon, K. A. 1994. A performance and failure analysis of SAPHIRE with a MEDLINE test collection. *Journal of the American Medical Informatics Association* 1(1):51–60.

Herskovic, J. R.; Tanaka, L. Y.; Hersh, W.; and Bernstam, E. V. 2007. A day in the life of PubMed: analysis of a typical day's query log. *Journal of the American Medical Informatics Association* 14:212–220.

Hoffmann, R., and Valencia, A. 2004. A gene network for navigating the literature. *Nature Genetics* 36(7):664.

Horn, F.; Lau, A. L.; and Cohen, F. E. 2004. Automated extraction of mutation data from the literature: application of MuteXt to G protein-coupled receptors and nuclear hormone receptors. *Bioinformatics* 20(4):557–568.

Hripcsak, G.; Bakken, S.; Stetson, P. D.; and Patel, V. L. 2003. Mining complex clinical data for patient safety research: a framework for event discovery. *Journal of Biomedical Informatics* 36(1–2):120–130.

Hu, Z.; Narayanaswami, M.; Ravikumar, K.; Vijay-Shanker, K.; and Wu, C. 2005. Literature mining and database annotation of protein phosphorylation using a rule-based system. *Bioinformatics* 21(11):2759–2765.

Huang, M.; Zhu, X.; Hao, Y.; Payan, D. G.; Qu, K.; and Li, M. 2004. Discovering patterns to extract protein–protein interactions from full texts. *Bioinformatics* 20(18):3604–12.

Humphrey, S. M.; Rogers, W. J.; Kilicoglu, H.; Demner-Fushman, D.; and Rindflesch, T. C. 2006. Word sense disambiguation by selecting the best semantic type based on journal descriptor indexing: Preliminary experiment. *Journal of the American Society for Information Science and Technology* 57(1):96–113.

Humphreys, B. L., and Lindberg, D. A. 1993. The UMLS project: making the conceptual connection between users and the information they need. *Bulletin of the Medical Library Association* 81(2):170–177.

Hunter, L., and Cohen, K. B. 2006. Biomedical language processing: what's beyond PubMed? *Molecular Cell* 21:589–594.

Hunter, L.; Lu, Z.; Firby, J.; Baumgartner Jr., W. A.; Johnson, H. L.; Ogren, P. V.; and Cohen, K. B. 2008. OpenDMAP: An open-source, ontology-driven concept analysis engine, with applications to capturing knowledge regarding protein transport, protein interactions and cell-specific gene expression. *BMC Bioinformatics* 9(78).

Hunter, L. E. 2009. *The processes of life: An introduction to molecular biology*. MIT Press.

Ide, N. C.; Loane, R. F.; and Demner-Fushman, D. 2007. Essie: A concept-based search engine for structured biomedical text. *Journal of the American Medical Informatics Association* 14:253–263.

Jackson, P., and Moulinier, I. 2002. *Natural language processing for online applications: text retrieval, extraction, and categorization*. John Benjamins Publishing Company.

Jacquemart, P., and Zweigenbaum, P. 2003. Towards a medical question-answering system: A feasibility study. In Baud, R.; Fieschi, M.; Beux, P. L.; and Ruch, P., eds., *The New Navigators: From Professionals to Patients*, volume 95 of *Actes Medical Informatics Europe, Studies in Health Technology and Informatics*, 463–468. Amsterdam: IOS Press.

Jaeschke, R.; Guyatt, G. H.; and Sackett, D. L. 1994. Users' guides to the medical literature. iii. how to use an article about a diagnostic test. b. what are the results and will they help me in caring for my patients? the evidence-based medicine working group. *The Journal of the American Medical Association* 271(9):703–707.

Jenssen, T.-K.; Lægreid, A.; Komorowski, J.; and Hovig, E. 2001. A literature network of human genes for high-throughput analysis of gene expression. *Nature Genetics* 28:21–28.

JFP. 2010. Clinical inquiries. The Journal of Family Practice. Available from: http://www.jfponline.com. accessed 2010.

Jiang, J., and Zhai, C. 2007. An empirical study of tokenization strategies for biomedical information retrieval. *Information Retrieval* 10(4-5):341–363.

Jimeno-Yepes, A., and Aronson, A. R. 2010. Knowledge-based biomedical word sense disambiguation: comparison of approaches. *BMC Bioinformatics* 11(5):569.

Jin, Y.; McDonald, R. T.; Lerman, K.; Mandel, M. A.; Carroll, S.; Liberman, M. Y.; Pereira, F. C.; Winters, R. S.; and White, P. S. 2006. Automated recognition of malignancy mentions in biomedical literature. *BMC Bioinformatics* 7.

Jin, F.; Huang, M.; Lu, Z.; and Zhu, X. 2009. Towards automatic generation of gene summary. In *Proceedings of the BioNLP 2009 Workshop*, 97–105. Boulder, Colorado: Association for Computational Linguistics.

Joachims, T. 1999. Making large-scale SVM learning practical. In Schölkopf, B.; Burges, C.; and Smola, A., eds., *Advances in kernel methods: Support vector learning*. MIT Press.

Johnson, D.; Zou, Q.; Dionisio, J.; Liu, V.; and Chu, W. 2002. Modeling medical content for automated summarization. *Annals of the New York Academy of Sciences* 980:247–258.

Johnson, H. L.; Cohen, K. B.; Baumgartner Jr., W. A.; Lu, Z.; Bada, M.; Kester, T.; Kim, H.; and Hunter, L. 2006. Evaluation of lexical methods for detecting relationships between concepts from multiple ontologies. *Pac Symp Biocomput*, 28–39.

Johnson, S. B. 1999. A semantic lexicon for medical language processing. *J Am Med Inform Assoc* 6(3):205–218.

Jurafsky, D., and Martin, J. H. 2008. *Speech and language processing: An introduction to natural language processing, computational linguistics, and speech recognition*. Pearson Prentice Hall.

Kan, M.-Y.; McKeown, K. R.; and Klavans, J. L. 2001a. Applying natural language generation to indicative summarization. In *Proceedings of the 8th European workshop on Natural Language Generation - Volume 8*, EWNLG '01, 1–9. Morristown, NJ, USA: Association for Computational Linguistics.

Kan, M.-Y.; McKeown, K. R.; and Klavans, J. L. 2001b. Domain-specific informative and indicative summarization for information retrieval. In *Proceedings of the Document Understanding Workshop (DUC 2001), New Orleans*.

Kaner, C.; Bach, J.; and Pettichord, B. 2002. *Lessons learned in software testing: a context-driven approach*. John Wiley and Sons, Inc.

Kaner, C.; Nguyen, H. Q.; and Falk, J. 1999. *Testing computer software, 2nd edition*. John Wiley and Sons.

Kann, M.; Ofran, Y.; Punta, M.; and Radivojac, P. 2006. Protein interactions and disease. In *Pacific Symposium on Biocomputing*, 351–353. World Scientific Publishing Company.

Katz, B.; Lin, J.; and Felshin, S. 2001. Gathering knowledge for a question answering system from heterogeneous information sources. In *Proceedings of the ACL 2001 Workshop on Human Language Technology and Knowledge Management*.

Kerrien, S.; Alam-Faruque, Y.; Aranda, B.; Bancarz, I.; Bridge, A.; Derow, C.; Dimmer, E.; Feuermann, M.; Friedrichsen, A.; Huntley, R.; Kohler, C.; Khadake, J.; Leroy, C.; Liban, A.; Lieftink, C.; Montecchi-Palazzi, L.; Orchard, S.; Risse, J.; Robbe, K.; Roechert, B.; Thorneycroft, D.; Zhang, Y.; Apweiler, R.; and Hermjakob, H. 2006. IntAct – open source resource for molecular interaction data. *Nucleic Acids Research* 35.

Kilicoglu, H., and Bergler, S. 2009. Syntactic dependency based heuristics for biological event extraction. In *BioNLP '09 Proceedings of the Workshop on Current Trends in Biomedical Natural Language Processing: Shared Task*, 119–127.

Kim, J.-D.; Ohta, T.; Tateisi, Y.; and Tsujii, J. 2003. Genia corpus – a semantically annotated corpus for bio-textmining. *Bioinformatics* 19(Suppl. 1):180–182.

Kim, J.-D.; Ohta, T.; Pyysalo, S.; Kano, Y.; and Tsujii, J. 2009. Overview of BioNLP'09 shared task on event extraction. In *BioNLP 2009 Companion Volume: Shared Task on Entity Extraction*, 1–9.

Kipper-Schuler, K. 2005. *VerbNet: A broad-coverage, comprehensive verb lexicon*. Ph.D. Dissertation, University of Pennsylvania dissertation.

Kogan, Y.; Collier, N.; Pakhomov, S.; and Krauthammer, M. 2005. Towards semantic role labeling & IE in the medical literature. In *AMIA 2005 Symposium Proceedings*, 410–414.

Krallinger, M.; Leitner, F.; Rodriguez-Penagos, C.; and Valencia, A. 2008. Overview of the protein–protein interaction annotation extraction task of BioCreative II. *Genome Biology* 9(Suppl. 2).

Krallinger, M.; Leitner, F.; and Valencia, A. 2007. Assessment of the second BioCreative PPI task: automatic extraction of protein–protein interactions. In *Proceedings of the Second BioCreative Challenge Evaluation Workshop*.

Kucera, H.; Francis, W. N.; and Carroll, J. B. 1967. *Computational analysis of present day American English*. Brown University Press.

Lancaster, F. W. 1969. Medlars: Report on the evaluation of its operating efficiency. *American Documentation* 20(2):119–148.

Laupacis, A.; Wells, G.; Richardson, W. S.; and Tugwell, P. 1994. Users' guides to the medical literature. v. how to use an article about prognosis. evidence-based medicine working group. *The Journal of the American Medical Association* 272(3):234–237.

Lesk, M. 1986. Automatic sense disambiguation using machine readable dictionaries: how to tell a pine cone from an ice cream cone. In *SIGDOC '86: Proceedings of the 5th annual international conference on systems documentation*, 24–26. New York, NY, USA: ACM Press.

Levine, M.; Walter, S.; Lee, H.; Haines, T.; Holbrook, A.; and Moyer, V. 1994. Users' guides to the medical literature. iv. how to use an article about harm. evidence-based medicine working group. *The Journal of the American Medical Association* 271(20):1615–1619.

Lin, J. 2009. Is searching full text more effective than searching abstracts? *BMC Bioinformatics* 10(46).

Liu, H.; Christiansen, T.; Baumgartner Jr., W. A.; and Verspoor, K. 2013. BioLemmatizer: a lemmatization tool for morphological processing of biomedical text. *Journal of Biomedical Semantics*, 3:3.

Lu, Z.; Cohen, B. K.; and Hunter, L. 2006. Finding GeneRIFs via Gene Ontology annotations. In *PSB 2006*, 52–63.

Lu, Z. 2007. *Text mining on GeneRIFs*. Ph.D. Dissertation, University of Colorado School of Medicine.

Marcus, M. P.; Marcinkiewicz, M. A.; and Santorini, B. 1993. Building a large annotated corpus of English: the Penn Treebank. *Computational Linguistics* 19(2):313–330.

McConnell, S. 2004. *Code complete*. Microsoft Press, 2nd edition.

McCray, A. T.; Burgun, A.; and Bodenreider, O. 2001. Aggregating UMLS semantic types for reducing conceptual complexity. In *Proceedings of 10th World Congress on Medical Informatics (MEDINFO 2001)*, 216–220.

Müller, H.-M.; Kenny, E. E.; and Sternberg, P. W. 2004. Textpresso: an ontology-based information retrieval and extraction system for biological literature. *PLoS Biol* 2(11):e309.

Miller, G. 2006. A scientist's nightmare: software problem leads to five retractions. *Science* 314:1856–1857.

Morgan, A. A.; Hirschman, L.; Colosimo, M.; Yeh, A. S.; and Colombe, J. B. 2004. Gene name identification and normalization using a model organism database. *J. Biomedical Informatics* 37(6):396–410.

Myers, G. 1979. *The art of software testing.* John Wiley and Sons.
Narayanaswamy, M.; Ravikumar, K. E.; and Shanker, V. K. 2005. Beyond the clause: extraction of phosphorylation information from medline abstracts. *Bioinformatics* 21(Suppl. 1).
Neves, M. L.; Carazo, J.-M.; and Pascual-Montano, A. 2010. Moara: a Java library for extracting and normalizing gene and protein mentions. *BMC Bioinformatics* 11.
Ng, S.-K. 2006. Integrating text mining with data mining. In Ananiadou, S., and McNaught, J., eds., *Text mining for biology and biomedicine.* Artech House Publishers.
Nielsen, J. 1989. Usability engineering at a discount. In *Proceedings of the third international conference on human-computer interaction*, 394–401.
Ogren, P.; Cohen, K.; and Hunter, L. 2005. Implications of compositionality in the Gene Ontology for its curation and usage. In *Pacific Symposium on Biocomputing*, 174–185.
OLDMEDLINE 2011. Oldmedline data. http://www.nlm.nih.gov/ databases/databases_oldmedline.html.
Olsson, F.; Eriksson, G.; Franzén, K.; Asker, L.; and Lidén, P. 2002. Notions of correctness when evaluating protein name taggers. In *Proceedings of the 19th international conference on computational linguistics (COLING 2002)*, 765–771.
Ono, T.; Hishigaki, H.; Tanigami, A.; and Takagi, T. 2001. Automated extraction of information on protein–protein interactions from the biological literature. *Bioinformatics* 17(2):60–67.
Palmer, M.; Kingsbury, P.; and Gildea, D. 2005. The Proposition Bank: an annotated corpus of semantic roles. *Computational Linguistics* 31(1):71–106.
Pedersen, T. 2008. Empiricism is not a matter of faith. *Comput. Linguist.* 34(3):465–470.
Pestian, J. P.; Brew, C.; Matykiewicz, P.; Hovermale, D.; Johnson, N.; Cohen, K. B.; and Duch, W. 2007. A shared task involving multi-label classification of clinical free text. In *Proceedings of BioNLP 2007.* Association for Computational Linguistics.
Pratt, A. W., and Pacak, M. G. 1969. Automated processing of medical English. In *Proceedings of the 1969 conference on Computational linguistics*, 1–23.
Pratt, W., and Yetisgen-Yildiz, M. 2003. A study of biomedical concept identification: MetaMap vs. people. In *Proceeding of the 2003 Annual Symposium of the American Medical Informatics Association (AMIA 2003)*, 529–533.
Pyysalo, S.; Ohta, T.; Kim, J.-D.; and Tsujii, J. 2009. Static relations: a piece in the biomedical information extraction puzzle. In *Proceedings of the BioNLP 2009 Workshop*, 1–9. Boulder, Colorado: Association for Computational Linguistics.
Regev, Y.; Finkelstein-Landau, M.; Feldman, R.; Gorodetsky, M.; Zheng, X.; Levy, S.; Charlab, R.; Lawrence, C.; Lippert, R. A.; Zhang, Q.; and Shatkay, H. 2002. Rule-based extraction of experimental evidence in the biomedical domain: the KDD cup 2002 (task 1). *SIGKDD Explor. Newsl.* 4(2):90–92.
Richardson, W. S., and Wilson, M. C. 1997. On questions, background and foreground. *Evidence Based Health Care Newsletter* 17:8–9.
Richardson, W. S.; Wilson, M. C.; Nishikawa, J.; and Hayward, R. S. 1995. The well-built clinical question: A key to evidence-based decisions. *American College of Physicians Journal Club* 123(3):A12–A13.
Rindflesch, T.; Tanabe, L.; Weinstein, J.; and Hunter, L. 2000. EDGAR: extraction of drugs, genes and relations from the biomedical literature. In *Pacific Symposium on Biocomputing*, 515–524.
Rosario, B., and Hearst, M. A. 2004. Classifying semantic relations in bioscience texts. In *Proceedings of ACL 2004*, 430–437.

Rosario, B., and Hearst, M. 2005. Multi-way Relation Classification: Application to Protein–protein Interactions. In *Proceedings of the HLT-NAACL*, volume 5.

Rosenberg, W., and Donald, A. 1995. Evidence based medicine: an approach to clinical problem-solving. *British Medical Journal* 310(6987):1122–1126.

Rzhetsky, A.; Iossifov, I.; Koike, T.; Krauthammer, M.; Kra, P.; Morris, M.; Yu, H.; Duboué, P. A.; Weng, W.; Wilbur, W. J.; Hatzivassiloglou, V.; and Friedman, C. 2004. Geneways: a system for extracting, analyzing, visualizing, and integrating molecular pathway data. *Journal of Biomedical Informatics* 37:43–53.

Sackett, D. L.; Straus, S. E.; Richardson, W. S.; Rosenberg, W.; and Haynes, R. B. 2000. *Evidence-Based Medicine: How to Practice and Teach EBM*. Edinburgh: Churchill Livingstone, second edition.

Saeed, M.; Villarroel, M.; Reisner, A. T.; Clifford, G.; Lehman, L.; Moody, G.; Heldt, T.; Kyaw, T. H.; Moody, B.; and Mark, R. G. 2011. Multiparameter intelligent monitoring in intensive care II (MIMIC-II): A public-access intensive care unit database. *Critical Care Medicine* 39(5):952–960.

Sandusky, R., and Tenopir, C. 2008. Finding and using journal article components: Impacts of disaggregation on teaching and research practice. *Joural of the American Society for Information Science and Technology* 59(6):970–982.

Schuemie, M. J.; Kors, J. A.; and Mons, B. 2005. Word sense disambiguation in the biomedical domain: an overview. *J Comput Biol* 12(5):554–565.

Schwartz, A., and Hearst, M. 2003. A simple algorithm for identifying abbreviation definitions in biomedical text. In *Pacific Symposium on Biocomputing*, volume 8, 451–462.

Settles, B. 2005. ABNER: an open source tool for automatically tagging genes, proteins, and other entity names in text. *Bioinformatics* 21(14):3191–3192.

Shah, P. K.; Jensen, L. J.; Boué, S.; and Bork, P. 2005. Extraction of transcript diversity from scientific literature. *PLoS Computational Biology* 1(1):67–73.

Shapiro, A. R. 1980. A system for conceptual analysis of medical practices. In *Proceedings of the Annual Symposium on Computer Application in Medical Care*, 867–872.

Shatkay, H.; Chen, N.; and Blostein, D. 2006. Integrating image data into biomedical text categorization. *Bioinformatics* 22(14):446–453.

Siadaty, M. S.; Shu, J.; and Knaus, W. A. 2007. Relemed: sentence-level search engine with relevance score for the MEDLINE database of biomedical articles. *BMC Medical Informatics and Decision Making* 7(1).

Sibanda, T., and Uzuner, O. 2006. Role of local context in automatic deidentification of ungrammatical, fragmented text. In *Proceedings of the Human Language Technology Conference of the NAACL, Main Conference*, 65–73. New York City, USA: Association for Computational Linguistics.

Smalheiser, N. R., and Swanson, D. R. 1999. Implicit text linkages between Medline records: Using Arrowsmith as an aid to scientific discovery. *LIBRARY TRENDS* 48(1):48–59.

Smith, R., and Chalmers, I. 2001. Britain's gift: a "medline" of synthesised evidence. *BMJ* 323:1437–1438.

Smith, R. 1996. What clinical information do doctors need? *BMJ* 313:1062–1068.

Srinivasan, P. 1996. Query expansion and MEDLINE. *Information Processing and Management* 32(4):431–443.

Stetson, P. D.; Johnson, S. B.; Scotch, M.; and Hripcsak, G. 2002. The sublanguage of cross-coverage. In *Proc. AMIA 2002 Annual Symposium*, 742–746.

Stevenson, M.; Guo, Y.; Gaizauskas, R.; and Martinez, D. 2008. Disambiguation of biomedical text using diverse sources of information. *BMC Bioinformatics* 9(Suppl 11):s7.

Sundheim, B. M. 1992. Overview of the fourth message understanding evaluation and conference. In *Proceedings of the 4th conference on Message understanding*, MUC4 '92, 3–21. Stroudsburg, PA, USA: Association for Computational Linguistics.

Swanson, D. R. 1960. Searching natural language text by computer. *Science* 132(3434):1099–1104.

Swanson, D. R. 1986a. Fish oil, Raynaud's syndrome, and undiscovered public knowledge. *Perspectives in Biology and Medicine* 30:7–18.

Swanson, D. R. 1986b. Undiscovered public knowledge. *Libr Q* 56(2):103–118.

Tanabe, L.; Scherf, U.; Smith, L. H.; Lee, J. K.; Hunter, L.; and Weinstein, J. N. 1999. MedMiner: an Internet text-mining tool for biomedical information, with application to gene expression profiling. *Biotechniques* 27(6):1210–1217.

Tateisi, Y.; Yakushiji, A.; Ohta, T.; and Tsujii, J. 2005. Syntax annotation for the GENIA corpus. In *Second international joint conference on natural language processing: Companion volume*, 220–225.

The Gene Ontology Consortium. 2000. Gene Ontology: tool for the unification of biology. *Nat Genet* 25(1):25–29.

Ting, K. M., and Witten, I. H. 1999. Issues in stacked generalization. *Journal of Artificial Intelligence Research* 10:271–289.

US-Congress. 1977. *Policy Implications of Medical Information Systems*. Washington, D.C.: OTA publications.

Uzuner, O.; South, B. R.; Shen, S.; and DuVall, S. L. 2011. 2010 i2b2VA challenge on concepts, assertions, and relations in clinical text. *Journal of the American Medical Informatics Association* 18:552–556.

Uzuner, O.; Luo, Y.; and Szolovits, P. 2007. Evaluating the state-of-the-art in automatic de-identification. *Journal of the American Medical Informatics Association* 14(5):550–563.

Varadan, R.; Assfalg, M.; Raasi, S.; Pickart, C.; and Fushman, D. 2005. Structural determinants for selective recognition of a Lys48-linked polyubiquitin chain by a UBA domain. *Molecular Cell* 18(6):687–698.

Verspoor, K.; Dvorkin, D.; Cohen, K. B.; and Hunter, L. 2009. Ontology quality assurance through analysis of term transformations. *Bioinformatics* 25(12):77–84.

Verspoor, K.; Cohen, K. B.; Lanfranchi, A.; Warner, C.; Johnson, H. L.; Roeder, C.; Choi, J. D.; Funk, C.; Malenkiy, Y.; Eckert, M.; Xue, N.; Baumgartner Jr., W. A.; Bada, M.; Palmer, M.; and Hunter, L. E. 2012. A corpus of full-text journal articles is a robust evaluation tool for revealing differences in performance of biomedical natural language processing tools. *BMC Bioinformatics* 13:207.

Verspoor, C.; Joslyn, C.; and Papcun, G. 2003. The Gene Ontology as a source of lexical semantic knowledge for a biological natural language processing application. In *Proceedings of the SIGIR'03 Workshop on Text Analysis and Search for Bioinformatics*.

Voorhees, E. M., and Harman, D. K. 2005. The Text REtrieval Conference. In Voorhees, E. M., and Harman, D. K., eds., *TREC: Experiment and evaluation in information retrieval*, 3–19. MIT Press.

Voorhees, E. M. 1999. *Natural language processing and information retrieval*. New York: Springer. 32–48. editor M T. Pazienza.

Wang, P.; Morgan, A. A.; Zhang, Q.; Sette, A.; and Peters, B. 2007. Automating document classification for the Immune Epitope Database. *BMC Bioinformatics* 8(269).

Wattarujeekrit, T.; Shah, P. K.; and Collier, N. 2004. PASBio: predicate-argument structures for event extraction in molecular biology. *BMC Bioinformatics* 5(155).

Weeber, M.; Mork, J.; and Aronson, A. 2001. Developing a test collection for biomedical word sense disambiguation. In *Proc AMIA Symp*, volume 746, 50.

Weizenbaum, J. 1966. Eliza – a computer program for the study of natural language communication between man and machine. *Commun. ACM* 9(1):36–45.

Wiegers, T. C.; Davis, A. P.; Cohen, K. B.; Hirschman, L.; and Mattingly, C. J. 2009. Text mining and manual curation of chemical-gene-disease networks for the Comparative Toxicogenomics Database (CTD). *BMC Bioinformatics* 10(326).

Wiegers, K. 2002. *Peer reviews in software: A practical guide*. Addison-Wesley.

Wilczynski, N.; McKibbon, K. A.; and Haynes, R. B. 2001. Enhancing retrieval of best evidence for health care from bibliographic databases: Calibration of the hand search of the literature. In *Proceedings of 10th World Congress on Medical Informatics (MEDINFO 2001)*, 390–393.

Xenarios, I.; Salwinski, L.; Duan, X. J.; Higney, P.; Kim, S.-M.; and Eisenberg, D. 2002. DIP, the Database of Interacting Proteins: A research tool for studying cellular networks of protein interactions. *Nucleic Acids Research* 30(1):303–305.

Xu, H.; Anderson, K.; Grann, V. R.; and Friedman, C. 2004. Facilitating cancer research using natural language processing of pathology reports. In *Studies in health technology and informatics*, 865–872.

Xu, R.; Supekar, K.; Morgan, A.; Das, A.; and Garber, A. 2008. Unsupervised method for automatic construction of a disease dictionary from a large free text collection. In *AMIA Annu Symp Proc*, 820–824.

Yang, X. F.; Su, J.; Zhou, G. D.; and Tan, C. L. 2004a. A NP-cluster based approach to coreference resolution. In *Proceedings of 20th International Conference on Computational Linguistics (COLING 2004)*, 226–232.

Yang, X.; Zhou, G.; Su, J.; and Tan, C. L. 2004b. Improving noun phrase coreference resolution by matching strings. In *IJCNLP04*, 326–333.

Yeh, A.; Morgan, A.; Colosimo, M.; and Hirschman, L. 2005. BioCreative task 1a: gene mention finding evaluation. *BMC Bioinformatics* 6(1).

Yuan, X.; Hu, Z.; Wu, H.; Torii, M.; Narayanaswami, M.; Ravikumar, K.; Vijay-Shanker, K.; and Wu, C. 2006. An online literature mining tool for protein phosphorylation. *Bioinformatics* 22(13):1668–1669.

Zhang, J.; Ga, J.; Zhou, M.; and Wang, J. 2001. Improving the effectiveness of information retrieval with clustering and fusion. *Computational Linguistics and Chinese Language Processing* 6(1):109–125.

Zieman, Y. L., and Bleich, H. L. 1997. Conceptual mapping of user's queries to medical subject headings. In *Proceedings of the 1997 Annual Symposium of the American Medical Informatics Association (AMIA 1997)*, 519–522.

Zou, Q.; Chu, W. W.; Morioka, C.; Leazer, G. H.; and Kangarloo, H. 2003. Indexfinder: A method of extracting key concepts from clinical texts for indexing. In *Proceedings of the 2003 Annual Symposium of the American Medical Informatics Association (AMIA 2003)*, 763–767.

Index